Petrophysics

Petrophysics:
A Practical Guide

Steve Cannon

WILEY Blackwell

This edition first published 2016 © 2016 by John Wiley & Sons, Ltd

Registered Office
John Wiley & Sons, Ltd, The Atrium, Southern Gate, Chichester, West Sussex, PO19 8SQ, UK

Editorial Offices
9600 Garsington Road, Oxford, OX4 2DQ, UK
The Atrium, Southern Gate, Chichester, West Sussex, PO19 8SQ, UK
1606 Golden Aspen Drive, Suites 103 and 104, Ames, Iowa 50010, USA

For details of our global editorial offices, for customer services and for information about how to apply for permission to reuse the copyright material in this book please see our website at www.wiley.com/wiley-blackwell.

The right of the author to be identified as the author of this work has been asserted in accordance with the UK Copyright, Designs and Patents Act 1988.

Library of Congress Cataloging-in-Publication Data

Cannon, Steve, 1955–
Petrophysics : a practical guide / by Steve Cannon.
     pages   cm
  Includes bibliographical references and index.
  ISBN 978-1-118-74674-5 (hardback) – ISBN 978-1-118-74673-8 (paper)
1. Petroleum–Geology.   2. Physical geology.   3. Geophysics.   I. Title.
  TN870.5.C328 2015
  553.2′8–dc23
                                                    2015022538

A catalogue record for this book is available from the British Library.

Wiley also publishes its books in a variety of electronic formats. Some content that appears in print may not be available in electronic books.

Cover image: An arbitrary cross section through a three dimensional porosity model of a reservoir built in Petrel (Schlumberger™) © Steve Cannon

Set in 10.5/12.5pt Times Ten by SPi Global, Pondicherry, India
Printed and bound in Singapore by Markono Print Media Pte Ltd

1   2016

Dedication

For Janet, as ever

Contents

Preface

This book has been written for those studying petroleum geology or engineering, for whom the role of the petrophysicist can become a lucrative and satisfying career. The handbook will be equally useful to students and practioners of environmental science and hydrogeology, where the understanding of groundwater flow is an important part of their technical remit. There is a comprehensive reference list included in the handbook that will cover some of the historic developments in petrophysics over the last 70 years; the book could bear the subtitle 'from Archie to anisotropy' and still include all the basic ideas captured in its pages. The handbook is subtitled 'a practical guide', and that is what I have set out to try and do: look at the pitfalls and obstructions encountered in any reservoir evaluation study and suggest alternative solutions and works-around.

It has also been developed in response to the needs of many younger colleagues who have not yet had the opportunities and experiences of an older generation of petroleum geologists and engineers. Most experienced petrophysicists in the industry have developed their own routines and preferred solutions to specific interpretation problems, and some of these different ideas will be reviewed for specific types of reservoir and the fluids they contain. Specialist petrophysicists and log analysts steeped in the details of wireline tool physics can sometimes lose the essential requirement of an interpretation, which is to provide usable input for some other static or dynamic model of a hydrocarbon reservoir.

I will quote one anecdote that may help explain what I mean. When learning how to use a new piece of log analysis software, I was afforded the opportunity by a colleague working for an oil company to test the software on a complete suite of wireline logs supported by core and sedimentological data, including XRD and SEM analysis. I had every tool then known to mankind available to employ on the usual problem of establishing porosity and water saturation in a well. I ran the log analysis software; the state-of-the-art software that used elemental analysis combined with a stochastic interpretation to generate the required results. My simple first-pass results were adequate, but no better than the previous deterministic solution I had generated. I was able to call upon the assistance of an expert log analyst in the service company with 20 years of experience in this field. I showed him the input data, the logs, the core analysis, XRD, etc., and my initial results; he felt that we could do better. After a couple of initial runs he said, 'I think I will relax the neutron a little', and my sandstone reservoir became a limestone; my comment was, 'it's a good job we have the core data; at least we know it is supposed to be sandstone!'. This was a life-changing event! I learnt that it is not enough to quality control the input data, you must quality control the output and make sure the results are sensible; often less experienced users of software solutions do not appreciate what the results should look like.

Until the 1980s, there were very few petrophysicists outside the oil company research laboratory: there were log analysts, core analysts, geologists and petroleum engineers, all of whom dabbled in the interpretation of wireline and core data. Log analysts had generally worked as field engineers with one of the many service companies, often being graduates in physics and electronics and thus well grounded in tool physics and data acquisition. It was with the development of computer-processed interpretations that log analysis became a tool of the many: often geologists with the basic knowledge to run the software but not always the experience to recognize bad data, the classic garbage-in, garbage-out syndrome.

There are also many oilfield service companies that provide wireline and/or LWD services from acquisition, through processing to interpretation. The biggest international companies are Baker Hughes, Halliburton, Schlumberger and Weatherford, and all provide services used in the evaluation of reservoirs; there are many other smaller and local companies that provide similar services. Throughout the book I have tried not to be biased towards one company or another; however, this is not always easy as some products or tools become associated with one or other company – my apologies should I appear to favour one organization over another, this is unintentional.

In writing this practical guide, I have started with the basic data acquisition and quality control of log and core data before moving into the actual interpretation workflow. Before starting an interpretation, however, it is crucial to establish a consistent database, and I give some suggestions on how this may be done, albeit this stage is often software dependent. The interpretation workflow follows a widely accepted series of steps from shale volume estimation, through porosity determination and finishes with the evaluation of water saturation. At each step I have presented a number of methods or techniques that may be selected depending on the available input data; these are not the only solutions, only the most common or simplest, so readers are invited to develop their own solutions for their own reservoirs. I have also tried to cover some of the more specialist log interpretation methods and their applications in reservoir characterization, especially how petrophysics links seismic data through a geological model to the dynamic world of reservoir simulation.

There are many people who have helped and guided me through my career, many no longer with us, so to them all I say, 'thank you'. I especially want to thank Roman Bobolecki, Andy Brickell, John Doveton, Jeff Hook, Mike Lovell, Dick Woodhouse and Paul Worthington, all proper petrophysicists! Finally, my thanks go to Andy Jagger and Nigel Collins of Terrasciences, who have supported me in many ways over the last 25 years, not least in providing a copy of T-Log for my use while writing this book.

Steve Cannon

1

Introduction

What is petrophysics? Petrophysics, as understood in the oil and gas industry, is the characterization and interaction of the rock and fluid properties of reservoirs and non-reservoirs:

1. determining the nature of an interconnected network of pore spaces – *porosity*;
2. the distribution of oil, water and gas in the pore spaces – *water saturation*; and
3. the potential for the fluids to flow through the network – *permeability*.

Petrophysical interpretation is fundamental to the much of the work on the subsurface carried out by geologists, geophysicists and reservoir engineers and drillers. To characterize the subsurface successfully requires physical samples, electrical, chemical, nuclear and magnetic measurements made through surface logging, coring and drilling and wireline tools (sondes). Terms such as 'formation evaluation' and 'log analysis' are often used to capture specific parts of the petrophysical workflow, but should not be seen as synonyms. 'Rock physics', which sounds as though it might be similar, is usually reserved for the study of the seismic properties of a reservoir; similar concepts apply but at larger scale.

The evaluation, analysis and interpretation of these petrophysical data is as much an art as a science, as it requires an understanding of geology, chemistry, physics, electronics, mechanics and drilling technology. At its simplest, petrophysics determines the porosity and water saturation of a reservoir, then estimates the permeability of the rock and the mobility of the fluids in place. The interpretation is dependent on the lithology of the rocks being evaluated, as sandstone, limestone, shale and any other potential hydrocarbon-bearing rocks all have differing characteristics. The acquisition and interpretation techniques applied in formation evaluation have been developed over the last

Petrophysics: A Practical Guide, First Edition. Steve Cannon.
© 2016 John Wiley & Sons, Ltd. Published 2016 by John Wiley & Sons, Ltd.

century primarily by the oil and gas industry, but the principles are equally relevant in coal mining, hydrogeology and environmental science. The type of data acquired is generic and can be used in a number of different analytical ways; indeed, as computing power and microelectronics have developed over the last 30 years, more high-resolution data can be collected and used for ever more detailed interpretation. However, measurements can be influenced by a number of variables, including the borehole environment; borehole diameter, temperature, pressure and drilling fluid, all affect the quality and type of data acquired. The reservoir rocks and the fluids therein can further affect the data quality and interpretation – a virtuous or viscous circle depending on how you look at it.

This book can be divided into two sections: first data acquisition and second interpretations, applications and workflow. This introductory chapter reviews the basics of petrophysics, including the confusing topics of measurement units, reservoir lithology, basic measurements and how the results may be used and the value of information and data management.

- *Chapter 2* reviews data acquisition in some detail, from drilling data to core analysis and wireline logs. I have not tried to give a detailed description of wireline tool technology, because I am not a physicist or electronics engineer; I refer you to the appropriate manufacturers' publications. In an appendix I have tried to collect basic tool information, but I would direct you to the third edition of The Geological Interpretation of Well Logs (Rider and Kennedy, 2011) for a full description and discussion of the range of logging tools available.
- *Chapter 3* discusses rock and fluid properties and what controls porosity, water saturation and permeability in the reservoir. Each property is defined and described and how the measurements are made, with a discussion of uncertainty.
- *Chapter 4* is focused on data quality control, especially the validation of log data and the integration with core data.
- *Chapter 5* looks at the characteristic response of different logs to reservoir rocks and fluids and how the data may be used in log analysis. The response to shales and matrix and fluid properties are fundamental.
- *Chapter 6* is about the evaluation of porosity and formation water resistivity and estimation of water saturation.
- *Chapter 7* looks at different petrophysical workflows, starting with data management and then quick-look single-well analyses, followed by multi-well studies. This part of the process is supported by worked examples.
- *Chapter 8* is called 'beyond log analysis' and looks at permeability estimation, cut-offs and zone averages, saturation height relationships, pressure measurements and fluid contacts. There is also a discussion of

lithology prediction, facies analysis and rock typing and also integration with seismic data.

- *Chapter 9* looks at carbonate reservoir characterization.
- *Chapter 10* describes the role of petrophysics in reservoir modelling, with a particular emphasis on property modelling in three dimensions.

One outcome of a petrophysical analysis forms the basis of the estimation of fluids in place, upon which, together with the gross rock volume of a reservoir, major investment decisions are made by oil and gas companies: the quality of the interpretation will change with time as new wells and new data are collected, so there is a need for consistency in approach at all times. One aspect that should never be forgotten is that most of the measurements that are made are a proxy for the real property that we are trying evaluate: porosity is never actually measured but interpreted from a density or neutron log; water saturation is interpreted from a resistivity measurement, dependent on the analyst knowing some fundamental properties of the formation fluid. A petrophysicist therefore has to be a general scientist with a strong numerical bias to be able to cut through the complex analytical methods and uncertainties inherent in the process of evaluating a reservoir; above all, a petrophysicist must be imaginative and thorough in their analysis and be flexible in their attitude to an interpretation that will change over time through either additional data or greater insight.

Beyond volumetric estimation, petrophysics is at the core of many other subsurface disciplines: the geophysicist relies on correctly edited and calibrated logs for depth conversion and rock property analysis, likewise the geologist for well correlation, reservoir modelling and fluid contact estimation, and the engineers for well completions and pressure prediction and as input for dynamic simulation. How you approach a petrophysical data set will often depend on the objective of the study: a single-well log analysis without core data requires a very different workflow to that adopted for a full-field petrophysical review.

Petrophysics is not just log analysis – it is log analysis within a geological context or framework, supported by adequate calibration data, including sedimentology, core analysis and dynamic data from pressure measurements and well tests (Figure 1.1). Logs do not measure porosity, permeability or water saturation; they make measurements of acoustic velocity, electrical conductivity and various nuclear relationships between the rock and the fluids to allow computer programs to process and interpret the results. The petrophysicist role is to validate and organize the input data and to understand and calibrate the results. A little harsh, you may say, but how many petrophysicists do the job without using log analysis software and how many integrate the analysis with the geological interpretation?

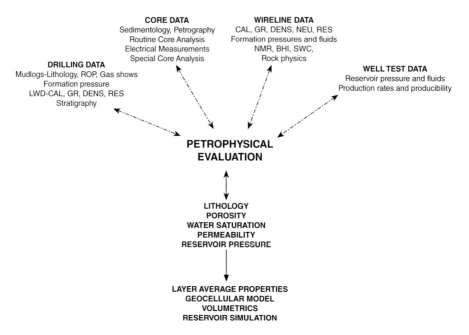

Figure 1.1 Petrophysical evaluation: schematic showing the primary data sources, products and deliverables of an integrated petrophysical evaluation.

1.1   The basics

It is worthwhile looking at the context in which the rest of the book lies before diving into the detail. Although not attempting to be a primer in geology, physics or chemistry, we will touch on these disciplines as we progress, so I will try to set the scene and leave the reader to dig deeper into interesting subject matter from the references. However, it is worth considering that both of our primary sources of data, wireline/LWD (logging while drilling) and core data, present challenges in terms of sampling, data quality and integration. Log measurements, although made *in situ*, are invariably indirect; we seldom measure an actual property of the rock, only one inferred from its response to physical input: core measurements are broadly speaking direct but they are *ex situ*. It is not my intention to describe in any detail the tool physics behind logging measurements, as there are many other books that cover this vital part of the technology; rather, this handbook is designed for the user of these data to evaluate the potential commercial value of a hydrocarbon reservoir.

All the log measurements that are made come from one or more penetrations of a reservoir made by a drill bit usually between 6 and 12½ inches in diameter, attached to a drill-string often several thousands of feet or metres long; we use this penetration to infer reservoir properties tens to thousands of metres away from the borehole (Figure 1.2). The borehole environment at depth is hostile;

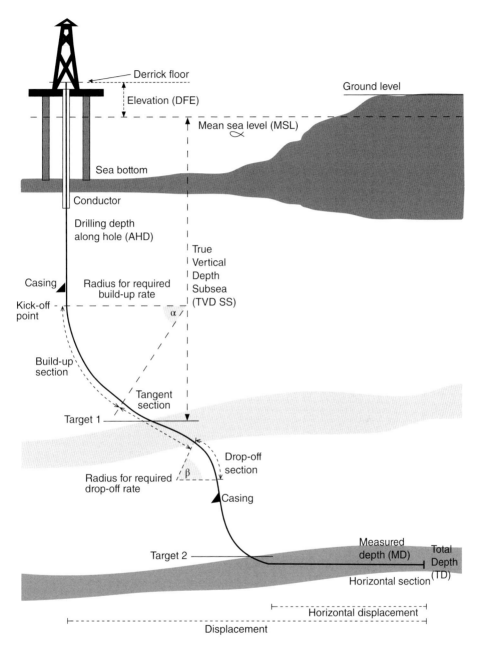

Figure 1.2   Depth measurement: terminology used to describe the stages and geometry of a well path designed to achieve a number of geological objectives.

Table 1.1   Comparison of different unit systems of measurement.

Measurement	SI units	Metric/imperial units	Field units	Abbreviation
Length/distance	Metre (m)	Metre/foot	Metre/foot	m/ft
Mass	Kilogram (kg)	Kilogram/pound	Pound	kg/lb
Time	Second (s)	Second	Second	s
Temperature	Kelvin (K)	Centigrade	Fahrenheit	°C/°F
Amount of substance	Mole (mol)	Mole	Parts per million	mol/ppm
Pressure	–	Pascal/bar	Pounds per square inch	Pa/bar
Volume	–	Cubic metre/barrel	Barrel	m³/bbl
Area	–	Hectare/acre	Acre	ha/ac

it can be hot enough to bake the sensitive electronics in the tools or be at pressures that result in the drilling mud being forced into the borehole wall (invasion) such that all the tool measures is a man-made fluid consisting of minerals and chemicals, which renders the results invalid or at best questionable. Even core measurements are made on material that has undergone physical change since it was cut; without careful handling, the change in confining pressure from reservoir to laboratory conditions will affect the pore volume and in-place fluids and even then measurement corrections are normally required to calibrate the results. Drillers, who generally do not like coring because of slow progress, have been heard to say that the only thing you know about a core once it has been cut is 'where it has come from, possibly'!

1.1.1   Units and abbreviations

The oil and gas industry can seem very confusing to the modern scientist brought up in the world of Système International (SI) units, because in general the industry uses either a mixed metric and 'imperial' unit system or 'field units' as the norm (Table 1.1).

The industry is also the home of more abbreviations and TLAs (three-letter acronyms) than probably any other, apart from the medical professions. There is a 'complete' glossary as an appendix; however, those given in Table 1.2 are some of the more pertinent for use in petrophysics.

1.1.2   Cores and logs

The two primary sources of reservoir information acquired during drilling of a well are cores and logs. Coring can be an expensive and time-consuming process that is usually reserved for potential reservoir sections. When the top reservoir is reached, signalled by a rapid increase in drilling rate and the presence of hydrocarbon shows, drilling is halted and the drill string recovered and the bit replaced

Table 1.2 Common abbreviations and three-letter acronyms.

Abbreviation	Meaning	Application
API	American Institute of Petroleum	Measure of gamma-ray activity; oil density
a	Archie exponent of tortuosity	Used in calculation of FRF and S_w
BHA	Bottom hole assembly	Drill-string from bit to top of drill collars
CAL	Calliper	Measures borehole diameter and rugosity
CPOR/CPERM	Core porosity/permeability	Core-derived porosity and permeability
DENS	Density log	Bulk density of formation from induced gamma activity
FRF	Formation resistivity factor	Core-derived resistivity of fully saturated sample
FVF	Formation volume factor	Ratio of oil volume at reservoir and surface conditions
GR/NGS	Gamma-ray log/spectral gamma log	Natural gamma radioactivity of formation
GDEN	Core grain density	Core-derived grain density of unsaturated sample
GRV	Gross rock volume	Volume of rock above a fixed datum
GIIP/STOIIP	Gas/stock tank oil initially in place	Hydrocarbons in place at time of discovery
LWD/MWD	Logging/measurement while drilling	Real-time telemetry and sensor measurements
m	Archie cementation exponent	Used in calculation of FRF and S_w
MW	Mud weight	Density of drilling fluid, usually in pounds per gallon, or specific gravity
NMR	Nuclear magnetic resonance	Uses the magnetic moment of hydrogen atoms to determine porosity and pore size distribution
n	Archie saturation exponent	Used in calculation of S_w from FRF
NTG	Net-to-gross ratio	Ratio of reservoir/pay to non-reservoir
NEUT	Neutron log	Measure of total hydrogen in a formation from water- and hydrocarbon-bearing pores
P_c	Capillary pressure	Fluid pressure/buoyancy of hydrocarbon–water systems
POR/PERM	Absolute porosity/permeability	Measure of connected pores
PhiT/PhiE	Porosity total/porosity effective	Porosity of isolated and connected pores
ROP	Rate of penetration	Drilling rate in feet or metres per hour
R_w	Resistivity of formation water	Function of water salinity
R_t	Formation resistivity	True resistivity of rock plus fluids
SONIC	Sonic log	Acoustic velocity of formation
S_w	Water saturation	Volume of water in pores
S_{wirr}	Irreducible water saturation	Volume of capillary-bound water/immoveable
S_{wc}	Connate water	Water trapped during deposition of sediments
Vugs/vuggy	Pore type in carbonates	Usually isolated or poorly connected pores

with a core barrel. Core barrels are usually made up of 30 ft lengths of pipe with a special coring head and retrieval mechanism, the catcher. There are in fact an inner and an outer barrel that can rotate independently; the inner barrel is the repository for the core as it is being cut. Upon retrieval at the surface, the core is stabilized and sent to shore for analysis; on occasion, some samples are evaluated at the well site, but this is becoming less and less common.

Logs are acquired while drilling (LWD) and also at the end of a hole-section on wireline. LWD and wireline logs represent among the most important data types available to a reservoir geoscientist or petrophysicist because they provide a continuous record of borehole measurements that can be used to interpret the environment of deposition of a sequence, the petrophysical properties and also the fluid distribution in the reservoir; in other words, to answer the questions do these rocks contain oil and gas and will it flow? However, the log measurements are greatly influenced by a number of variables, including the borehole environment, the rocks themselves and the type of fluid used to drill the well.

1.1.3 Lithology identification

Most hydrocarbons are found in either clastic or carbonate reservoirs; clastic rocks such as sandstones comprise grains of quartz, feldspar, mica, lithic fragments, clays and exotic minerals. Depending on the sediment source, these grains will be deposited in different proportions and represent different depositional processes, and these distinctive characteristics should be discernible to some extent in the petrophysical data collected. To a petrophysicist, clastics are either sand or shale or maybe siltstone; sometimes a grain size distinction such as coarse or fine may be added, or whether shale has a high organic content. For a geologist, a much greater variation in sandstone classification is required based on a simple ternary diagram with quartz, feldspar and lithic (QFL) fragments at the apices (Figure 1.3). By strict definition, shale is a fine-grained clastic rock composed of mud comprising clay minerals and silt grains or other minerals, mainly quartz, feldspar and carbonate, that exhibits a fissile nature; it splits along lamina. A mudstone does not show this fissility, but in composition may be exactly the same as the equivalent shale. For a complete review of sandstone petrology, see Folk (1980).

Carbonate rocks generally fall into three types for the petrophysicist: limestone, dolomite and anhydrite or evaporites, if a mixture of different salt deposits is recognized. The range of carbonate rocks to the geologist is even more extensive than the clastics; fortunately, many carbonate classification systems have some basis in pore types, giving a direct link to the petrophysical world (Figure 1.4). The petrophysics of carbonate reservoirs is a specialist role – highly challenging but ultimately very rewarding, especially when one considers the proportion of the world's oil to be found in these reservoirs. For an extensive study of carbonate reservoir characterization, see Lucia (1999).

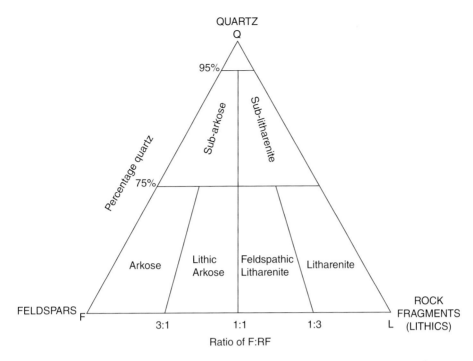

Figure 1.3 QFL plot: a standard lithology ternary plot based on the proportions of quartz, feldspar and rock fragments in sandstone. Source: after Folk (1980).

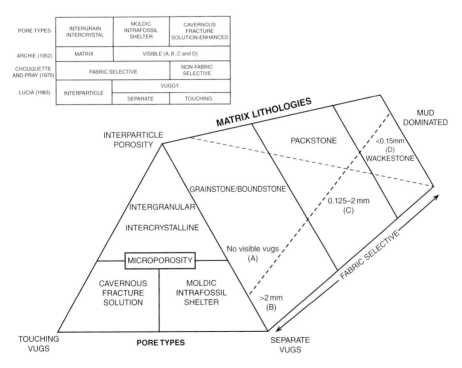

Figure 1.4 Carbonate pore types: classification of carbonate rock into intergranular and vuggy pore types; comparison of alternative classification schemes. Source: after Lucia (1999).

With the rapid increase in development of unconventional reservoirs in recent years, especially in North America, has come a series of new challenges for the petrophysicist: to evaluate their potential as both source rock and reservoir. Most of this book will consider conventional reservoirs; however, there will be some discussion of 'unconventionals' as appropriate. It is worth pointing out that the terms clay and shale are often used as though they mean the same thing: this is not the case and the differences will be discussed later.

1.1.4   Rock properties

The presence of an effective pore network and the capacity of it to allow fluids to flow through it are a function of a rock's primary depositional process, the resulting grain size distribution and the effect of post-depositional processes, principally compaction, chemical diagenesis and fracturing. This statement applies equally to clastic or carbonate reservoirs; however, the effects of post-depositional processes are generally more significant in carbonates. Clastic reservoirs can be unconsolidated or consolidated or lithified to varying extents depending on the post-depositional history of the sediments, the process of compaction and cementation. The degree of lithification can be obvious in some logs, such as an acoustic log, where the transit time of the sound waves will vary from slow to fast depending on the consolidation of the rock. This in turn may have an effect on the porosity of the rock; softer rocks generally have higher porosity.

1.1.5   Physics of a reservoir

The pore spaces of a hydrocarbon reservoir begin life filled with water that is either mobile or bound by capillary pressure. The water becomes displaced by hydrocarbons during migration because of the contrast in fluid density: water is more dense than either gas or oil; this is known as the drainage cycle. Under the correct structural or stratigraphic conditions, the hydrocarbons become trapped and continue to displace the water, until only the smallest pores remain water filled; this is connate water. Where water saturation is 100% and capillary pressure is zero is called the free water level (FWL), a datum that is defined by the physics of the reservoir (Figure 1.5). Where the reservoir is homogeneous and has large pore throats, the hydrocarbon water contact and FWL will be contiguous, but if the reservoir quality is poor at the base of the hydrocarbon column, the two levels will be separated by a transition zone.

1.1.6   Porosity

Porosity is defined as the capacity of a rock to store fluids and estimated as the ratio of the pore volume to the bulk volume. Porosity is a non-dimensional parameter expressed as a fraction or percentage. The porosity of a rock comprises

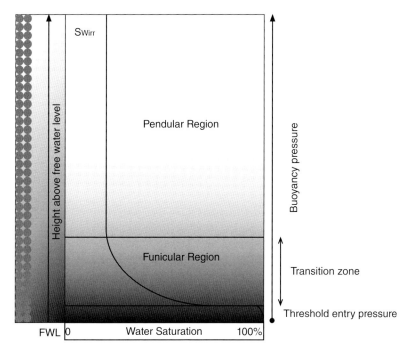

Figure 1.5 Physics of the reservoir: representation of fluid distribution within an oil reservoir based on the relationship between water saturation, capillary pressure and the free water level datum.

two main elements, primary depositional or intergranular porosity and secondary porosity, which may be the result of grain or particle dissolution or present as microporosity in authigenic clays (Figure 1.6a). Porosity may be defined as *effective* or *total* depending on whether it includes porosity associated with clays; some tools measure total porosity and must be corrected for the clay content. This is a simple classification that does not include all carbonate rocks or certain clay-rich shale reservoirs. Fractured reservoirs need also to be treated separately, being defined as having a dual porosity system, matrix and fracture.

1.1.7 Water saturation

Water saturation (S_w) is the proportion of total pore volume occupied by formation water; hydrocarbon saturation is derived from the relationship $S_h = 1 - S_w$. It may be expressed as a fraction or a percentage depending on how porosity is defined (Figure 1.6b). Another direct link to porosity terminology exists, as water saturation can be either a total or an effective value. Logs measure both the mobile water and the clay-bound water in the pore space. The terms irreducible, residual, connate and initial water saturation are also commonly used, sometimes without due regard to the meaning.

(a)

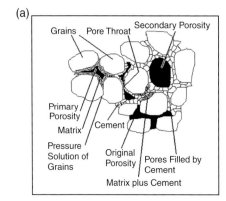

Porosity = volume of pore space
         ───────────────────────
         total volume of rock

Expressed as a fraction or percentage
A function of grain size and packing
Can be expressed as TOTAL or EFFECTIVE

Primary porosity reduces with compaction due to burial and lithification/cementation
Secondary porosity is a result of disolution of unstable minerals.

(b)

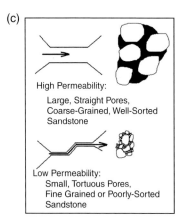

Water saturation (S_w) = pore volume filled with water
                         ──────────────────────────────
                         total pore volume

Hydrocarbon saturation (S_h) = $(1 - S_w)$

Can be expressed as TOTAL or EFFECTIVE property dependent
Be consistent in use of terms

(c)

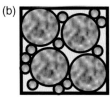

High Permeability:

Large, Straight Pores,
Coarse-Grained, Well-Sorted
Sandstone

Low Permeability:
Small, Tortuous Pores,
Fine Grained or Poorly-Sorted
Sandstone

Measure of the ability of a reservoir to conduct fluids
A dynamic property dependent on rock and fluid
characteristics with a directional (vector) component
May be a predictable relationship with porosity

$$Q = \frac{kA\Delta p}{\mu L}$$

Absolute, effective and relative permeability values may
be required as a key input for dynamic simulation

Figure 1.6   (a) Porosity: the relationship between volume of pore space and total volume of rock is a function of grain size, sorting and packing at time of deposition. Post-depositional processes such as compaction and diagenesis can alter the original relationship. (b) Water saturation: the proportion of the total reservoir pore volume filled with water: the remaining pore volume is filled with oil or gas, not necessarily hydrocarbon gas. (c) Permeability: the ability of a reservoir to conduct fluids through an interconnected pore network.

Irreducible water saturation (S_{wirr}) is defined as the minimum S_w at high capillary pressure and saturation, as the effective permeability to water approaches zero. The initial water saturation (S_{wi}) is the proportion of water in the reservoir at the time of discovery and may be synonymous with connate water, the water saturation at time of deposition, if no hydrocarbons are present. In a hydrocarbon-bearing reservoir, S_{wirr} is always less than S_{wi}. The term 'transition zone' also has more than one meaning depending on who is using it: to a geologist or petrophysicist it is the zone between the lowest level of irreducible water and the free water level – this is a static definition; to a reservoir engineer it is an interval in a well that flows both oil or gas and water at the same time – the two 'zones' may be contiguous.

1.1.8 Permeability

Permeability (K or k) is the measure of the capacity of a reservoir to conduct fluids or for flow to take place between the reservoir and a wellbore. A dynamic property, permeability is dependent on the associated rock and fluid properties (Figure 1.6c); it is also one of the most difficult to measure and evaluate without data at all relevant scales – core, log and production test. At the microscopic or plug scale, permeability is a function of pore network and whether there are large or small pore throats and whether the connecting pathways are straight or tortuous; a function of grain size and sorting. Permeability is also a vector property as it may have a directional component, resulting in anisotropy. Permeability may vary greatly between the horizontal and vertical directions, impacting on the directional flow capacity of a reservoir. Given the difficulties in reliably measuring permeability, a qualitative assessment is often made depending on the hydrocarbon in place (Table 1.3).

Permeability is measured in darcies (D) but usually reported as millidarcies (mD), named after the French water engineer Henry Darcy, who first attempted to measure the flow of water through a vertical pipe packed with sand. The rate of flow (Q) is a function of the area (A) and length (L) of the pipe, the viscosity of the fluid (μ) and the pressure differential (Δp) between the ends of the pipe (Figure 1.6c). This law only applies to a single fluid phase and may be termed absolute or intrinsic permeability. Effective permeability (K_{eff}) is the permeability of one liquid phase to flow in the presence of another; relative permeability (K_r) is the ratio of effective to absolute permeability for a given saturation of the flowing liquid, i.e. permeability of oil in the presence of water (K_{ro}). Permeability is a key input for numerical reservoir simulation.

Relative permeability is the normalized value of effective permeability for a fluid to the absolute permeability of the rock. Relative permeability

Table 1.3 Permeability ranges for different qualitative descriptions of permeability.

Poor	<1 mD	'Tight' for gas
Fair	1–10 mD	'Tight' for oil
Moderate	10–50 mD	
Good	50–250 mD	
Excellent	>250 mD	

expresses the relative contribution of each liquid phase to the total flow
capacity of the rock.

1.1.9 Capillary pressure

Capillary pressure acts at a microscopic scale in the reservoir, which in
conjunction with viscous and gravitational forces define how a reservoir
performs dynamically. Capillary pressure occurs whenever two immiscible
fluids occur in the pore space of a rock and is defined as the pressure
difference measurable in the two phases (Figure 1.7a). There is an inherent
relationship between capillary pressure and water saturation because water
is retained in the pore space by capillary forces. Capillary pressure also
determines the fluid distribution and saturation in a reservoir, hence the
link to wettability.

1.1.10 Wettability

Wettability is a measure of a rock's propensity to adsorb water or oil molecules
on to its surface in the presence of the other immiscible fluid. At deposition, a
thin film of water is usually formed around the grains, leaving the rock water

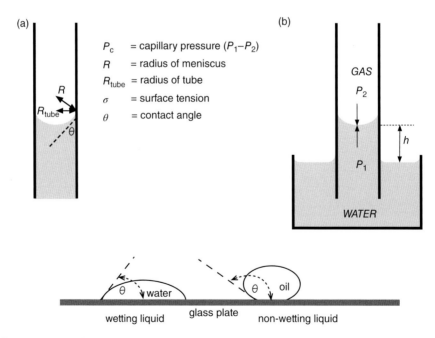

Figure 1.7 Capillary pressure (P_c) and wettability: (a) representation of a liquid-filled capillary
tube and the relationship between the buoyancy pressure generated between two immiscible
fluids; (b) the difference between wetting and non-wetting liquids as a function of the surface
tension and contact angle.

wet – the normal situation; however, carbonate rocks are commonly oil wet or have intermediate wettability. Wettability is a function of the surface tension between the solid grain and the fluid in the pores (Figure 1.7b).

It is important to understand the impact of wettability on the other dynamic properties of a rock as it controls the fluid saturation and distribution in a reservoir. Although most (clastic) reservoirs would be considered to be water wet, under certain conditions all reservoirs can become oil wet, at least in part. Carbonate reservoirs have a greater tendency for the oil wet state because of the greater adsorption capacity of calcium/magnesium carbonate. Many reservoirs are of mixed wettability – oil wet in the large open pores and water wet in the smaller isolated pores often filled with microporous clays.

1.2 The results

When the petrophysical analysis of a single well or group of wells has been completed, the results will be used in a number of ways: as the estimate of hydrocarbon pay in a well; in making the decision to production test an exploration well; as input for a simple volumetric calculation; to build a 3D property model of a field; or as a key element in a major investment or divestment decision.

1.2.1 Hydrocarbon pay

The *gross reservoir thickness* is described as the total reservoir thickness in a well; layers of non-reservoir (shale) are discounted from the total, leaving a *net reservoir thickness*. The ratio is thus the *net-to-gross* (NTG). Hydrocarbon-bearing levels, *net pay*, in a well are normally defined in terms of minimum porosity and saturation values calculated from the log analysis; sometimes a permeability limit is also applied (Figure 1.8) This approach can be too harsh if the wrong cut-off parameters are applied. If sufficient net pay is recognized in a well, then the decision to test the interval or intervals is made; the results should be integrated into the rest of the reservoir evaluation.

1.2.2 Simple volumetrics

Map-based volumetrics require a *gross rock volume* (GRV), usually a top reservoir surface and hydrocarbon–water contact; an alternative is to use a simple slab model (Figure 1.9). Apart from a value for the formation volume factor (FVF), all the terms can be derived from log analysis.

$$\text{HIIP} = \text{GRV} \times \text{NTG} \times \Phi \times (1 - S_w) \times \text{FVF}$$

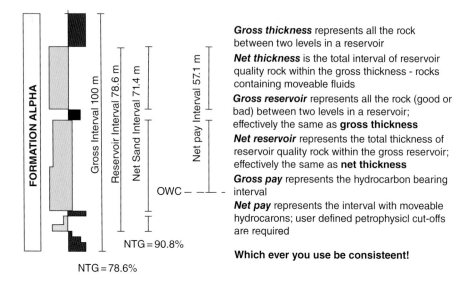

FORMATION ALPHA
Gross Interval 100 m
Reservoir Interval 78.6 m
Net Sand Interval 71.4 m
Net pay Interval 57.1 m
OWC
NTG = 90.8%
NTG = 78.6%
Gross thickness represents all the rock between two levels in a reservoir
Net thickness is the total interval of reservoir quality rock within the gross thickness - rocks containing moveable fluids
Gross reservoir represents all the rock (good or bad) between two levels in a reservoir; effectively the same as gross thickness
Net reservoir represents the total thickness of reservoir quality rock within the gross reservoir; effectively the same as net thickness
Gross pay represents the hydrocarbon bearing interval
Net pay represents the interval with moveable hydrocarons; user defined petrophysicl cut-offs are required
Which ever you use be consisteent!

Figure 1.8 Net to gross: terminology used to describe the proportions of an oil or gas reservoir in terms the different interval thicknesses.

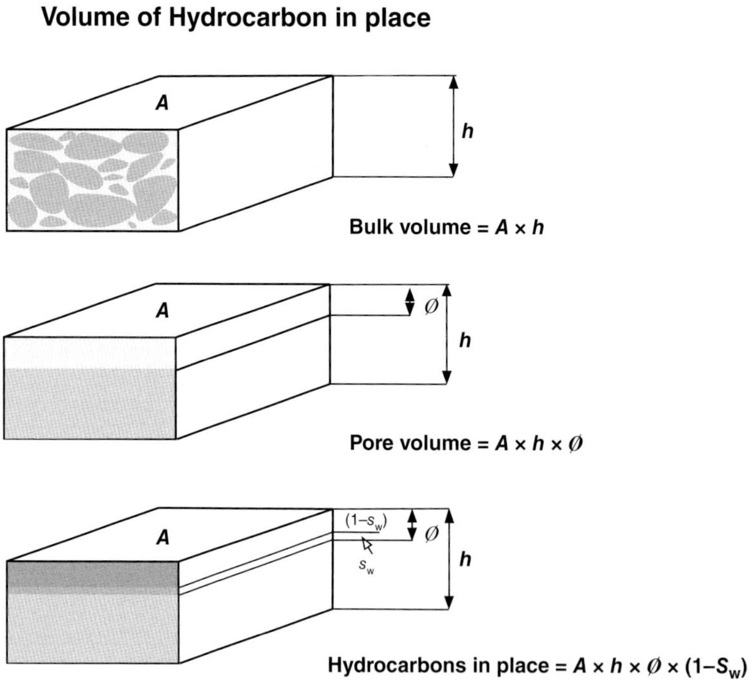

Volume of Hydrocarbon in place
A
h
Bulk volume = A x h
A
Ø
h
Pore volume = A x h x Ø
A
(1–Sw)
Sw
Ø
h
Hydrocarbons in place = A x h x Ø x (1–Sw)

Figure 1.9 Volume of HIIP: schematic to show the calculation of the volume of hydrocarbons in place in an oil or gas reservoir; to estimate potential resources it is necessary to apply the appropriate conversion factor from reservoir volume to surface volume, the formation volume factor.

Often these same data are run through a Monte Carlo analysis to give a range of results; specific cases can then be selected from a representative distribution.

1.2.3 3D static models

The most accurate volumetrics can be calculated in a 3D static or geocellular model; this is because it will be geometrically more correct, especially where the field is faulted. The inputs remain the same, although now the petrophysical properties are distributed throughout the model randomly or following some established trend, and uncertainties can be modelled to establish volumetric ranges for the field. These should always be compared with the simple Monte-Carlo results above to provide a sense check. As a rule of thumb, variations in *hydrocarbon initially in place* (HIIP) greater than ~25% are a function of the GRV; petrophysical properties properly constrained and distributed seldom impact the result by more than a few per cent.

1.2.4 Value of information

Acquiring petrophysical data, be they logs or core, is expensive; the formation evaluation programme for a well can be several hundred thousand dollars or can run into millions. The value of the information must always be seen in terms of where you are in the project cycle and also the use to which the data will be put. Sometimes too much data can result in indecision just as much as an incomplete dataset, and when dealing with brownfield developments where only an older log set is available, it may be more important to calibrate the basic data than to run the most sophisticated tools.

The key decision making data sets are as follows:

- Logs to establish lithology, porosity and fluids.
- Cores to confirm lithology and calibrate the log-derived properties and to establish a depositional environment; in an exploration well, cores may be replaced by image and scanner tools, but cores should be acquired in subsequent appraisal wells.
- Pressure measurements and fluid samples.
- Production data to establish that the reservoir will flow, recover fluids and test the limits of the hydrocarbons connected to the well.

With a good, well-distributed, basic data set, a robust reservoir description will be possible even if some of the 'bells and whistles' beloved of specialist disciplines are not available; any additional information should not be ignored but recognized as single points in the overall population and should not bias the interpretation.

1.3 Summary

In this chapter, we have looked at the basics of petrophysics; the types of data required for a petrophysical interpretation and some of the fundamental results of the analysis. It should be apparent that for an accurate calculation of porosity and water saturation, the petrophysicist is dependent on many input properties over which he or she has little control. Most of the measurements are either indirect or *ex situ* and are subject to sample bias, acquisition issues and experimental (human) error. The uncertainties are compounded when the results are used inappropriately or without sufficient caveats regarding how the data were collected, edited, manipulated or applied. In the simplest form, the results of a petrophysical study can be managed by applying cut-offs or error bounds, but when incorporated in some geostatistical model or multi-scale dynamic model, their application may be incorrect or inappropriate.

2

Data Acquisition

While the primary purpose of drilling an oil or gas well is to discover commercial hydrocarbons, often it is the acquisition of data that is the more important outcome of the process. Data are acquired both during drilling and after it has been completed, either upon reaching the terminal or total depth of a well or when a section of the borehole has been completed. There are four areas of data acquisition that most concern petrophysicists: drilling data, core analysis, wireline logging and well test data.

2.1 Drilling data

Data acquired during drilling can be summarized as physical, such as drill cuttings and gas records, and process information, such as rate of penetration (ROP), mud weight (MW) or formation pressure and real-time telemetry and sensor measurements (LWD). The physical data are recorded on the 'mudlog' (Figure 2.1) and the LWD measurements as a digital output of the various tools. Basic formation evaluation can be made with these data alone and may form the only record of some exploratory wells; development wells may also have very limited data acquisition programmes. However, a lot of information can be extracted from this basic data set, including lithology, presence of hydrocarbons and porosity and water saturation – in other words, an evaluation of the drilled formations. Physical samples such as the drill cuttings and gas samples also form the basis of more detailed studies on formation stratigraphy, source rock geochemistry and gas component analysis.

Mudlog data are 'lagged': the data reach the surface sometime after being drilled as the physical samples are pumped to the surface with the rest of the drilling mud. This 'lag-time' can easily be 45–60 min in deeper wells, and during the process the samples experience a degree of mixing. Combined with the fact that cuttings samples are usually only collected every 10 ft in the deeper sections of a well, this means that the record of lithology is a composite: reference

Petrophysics: A Practical Guide, First Edition. Steve Cannon.
© 2016 John Wiley & Sons, Ltd. Published 2016 by John Wiley & Sons, Ltd.

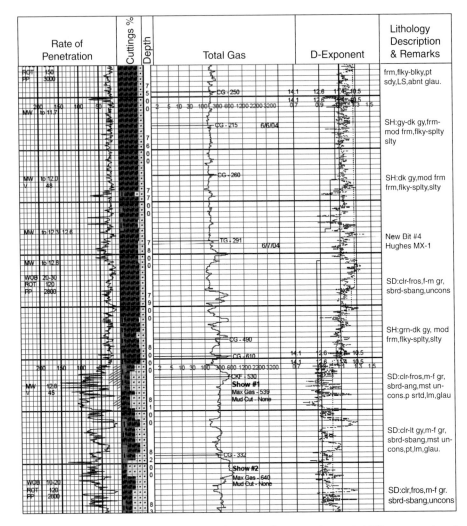

Figure 2.1 An example of a mudlog showing rate of penetration and drilling parameters in the first column, cuttings percentages and depth in columns 2 and 3, total gas in column 4 and the drilling exponent related to formation pressure in column 5. (For a colour version see Plates section).

to the ROP is required for sudden changes in lithology such as from shale to sand, termed a drilling break, where the ROP increases almost instantaneously. The physical drilling data, ROP, torque and weight-on-bit are real-time information seen on the drill floor, allowing the driller to optimize progress in different formations.

Gas readings are made as a continuous stream of data by several instruments, including a total gas analyser, a gas chromatograph and an H_2S detector.

Together these instruments provide an indication of increasing hydrocarbon and non-hydrocarbon gases present in the return mud flow; an increase in total gas combined with evidence of heavier gas components, C_1–C_5, may indicate the presence of mobile hydrocarbons in the formation.

LWD measurements are taken by self-contained tools near the drill bit as part of the bottom hole assembly (BHA). The data are recorded downwards (as the well is deepened) rather than upwards from the bottom of the hole (as wireline log data are recorded). LWD records are measured against time-while-drilling and then processed to convert the readings to depth. The position of the tools away from the bit may limit their use and effectiveness in some situations, and planning the type and order of the assembly can become critical. Operational or drilling requirements will influence the location of the LWD toolstring, especially when geo-steering or trying to maintain a high penetration rate. Data may be stored in the tool's memory during drilling or transmitted as pressure pulses in the mud column in real time. Typically, both modes will be used, with memory data being retrieved at the surface; these are the data that are considered the most reliable.

LWD, although sometimes unreliable and expensive, has the advantage of measuring properties of a formation before drilling fluids invade deeply. Common LWD measurements made include standard telemetry data (azimuth and inclination) and gamma ray, density, neutron porosity and resistivity data. Many boreholes prove to be difficult or even impossible to measure with conventional wireline tools, especially highly deviated wells. In these situations, the LWD measurement ensures that some measurement of the subsurface is captured in the event that wireline operations are not possible. Most types of LWD tools make measurements comparable to those made by wireline; however, there is seldom a direct equivalence in the actual values measured because the tool technology and processing are generally different. Where both measurements are available, the wireline results should be considered definitive and used to calibrate the LWD responses such that in later field life only LWD measurements need be made.

2.2 Coring and core analysis

Coring is the process of acquiring larger borehole samples, either as the hole is deepened using a specialized core bit and assembly (Figure 2.2) or as sidewall core samples acquired on wireline after the hole has been drilled. Cores and core samples are used to understand better the depositional environment of the formation and also to obtain measurements of porosity and permeability from potential reservoir units. Petrographic thin sections of core samples are used for mineralogical, textural and diagenetic studies and may include SEM images of pore size distribution. Selected core samples, usually representative

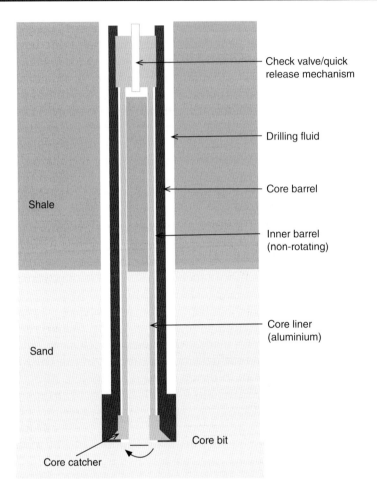

Figure 2.2   Schematic diagram of a coring assembly and barrel prior to retrieval.

of different facies, are used for electrical measurements to obtain values of a, m and n in the Archie equations to establish formation resistivity and estimate water saturation from logs. Core samples are also used to establish dynamic properties of the reservoir such as wettability, capillary pressure and relative permeability.

When a core is cut and recovered to the surface, it is no longer truly representative of the formation: it is no longer at the formation temperature or pressure and the fluids once contained in the pores will have expanded and either moved or evaporated as a result. Physical changes to the rock fabric can also be expected as grains or particles become disaggregated, affecting the porosity and permeability; having said that, starting a petrophysical study without core data is bound to result in more questions than answers. A continuous core through a reservoir provides the necessary geological information to describe the lithology, detrital and authigenic mineralogy, pore structure and the likely depositional setting and diagenetic history of the sequence; all of this information is essential for 'calibrating' the wireline logs during the core–log integration process.

Standard coring and core handling procedures result in much better recovery of cores today, leaving less opportunity for material to become misplaced before it can be correctly measured and sampled. Once the laboratory has received a core, it is laid out, measured and marked up and a quick description of the surface features, especially fractures, is made, prior to gamma logging and sampling. Sampling for routine core analysis should be done at regular intervals throughout the core, regardless of the lithology, as this will minimize sample bias; a core data set comprising only high-porosity or -permeability material probably will not represent the reservoir. Sampling for special core analysis experiments should be done by the geologist and petrophysicist together, to ensure that samples representative of the different facies or rock types are selected. These samples will often be whole pieces of core and a description of the sample should be made and a photograph taken, especially of any sedimentological contacts, before removal and preservation. The core is usually then cut (slabbed) into 1/3–2/3 sections before being set in a plastic resin, a process designed to preserve the core for the future. The slabs are photographed under white light and ultraviolet settings and then prepared for storage. A detailed sedimentological core description should be made at this time.

Routine sampling involves cutting 1–1½ inch diameter plugs at right-angles to the core surface every 1 ft or 25 cm with a water-cooled diamond and tungsten drill bit attached to a variable-speed, bench-mounted drilling rig. The samples are labelled prior to cleaning in a cool solvent chamber to remove liquid hydrocarbons and water. The process can take several days depending on the viscosity of the fluids and the connectivity of the pores and is not always completely successful. The samples are then dried in a carefully controlled humidity oven to remove any remaining water or solvent. This process, although carefully controlled, has the potential to damage any filamentous clay particles, such as illite, in the pores, and although not affecting porosity to any great extent it may alter permeability. A detailed petrographic analysis of the samples can indicate whether there has been any damage to the samples.

There are two groups of special core analysis experiments that influence a petrophysical study: the electrical measurements needed for the calculation of water saturation from logs and dynamic measurements required for reservoir modelling and flow prediction. Both types of experiments are time consuming and are done on only a few selected samples from a core or reservoir because of the cost involved; however, without these data, the uncertainties associated with reservoir characterization increase significantly.

Samples are usually of larger diameter plugs or even whole core pieces that have been thoroughly cleaned and validated as homogeneous and representative of a single rock type; internal integrity can be assessed with CT scanning methods.

2.3 Wireline logging

Wireline logs represent one of the most important data types available to a reservoir geoscientist or petrophysicist because they provide a continuous record of borehole measurements that can be used to interpret the environment of deposition of a reservoir, the petrophysical properties and also the fluid distribution in the reservoir; in other words, do these rocks contain oil and gas and will it flow? However, the log measurements are greatly influenced by a number of variables, including the borehole environment, the rocks themselves and the type of fluid used to drill the well. Digital log data are often provided in standard LAS 2.0 (log-ASCII) format developed by the Canadian Society of Professional Well Log Analysts (CSPWLA). This format records the tool set-up, log header information from each log run and also the continuous log data at an interval spacing of 6 in or 15 cm; other common formats you may come across are LIS and D-LIS.

The Schlumberger Oilfield Glossary (www.slb.com/glossary.aspx) defines a wireline log as 'a continuous measurement of formation properties with electrically powered instruments to infer properties and make decisions about drilling and production operations'. The log is in fact the record of the measurements, typically a long strip of paper. Measurements include electrical properties (resistivity and conductivities at various frequencies), acoustic properties, active and passive nuclear measurements, dimensional measurements of the wellbore, formation fluid sampling, formation pressure measurement, wireline-conveyed sidewall coring tools and others. In making wireline measurements, the logging tool (or sonde) is lowered into the open wellbore on a multiple conductor, contra-helically armoured wireline. The cable is wrapped around a motorized drum positioned near the rig floor that is guided manually during the logging run (Figure 2.3).

Once lowered to the bottom of the interval of interest, the measurements are taken on the way out of the wellbore. This is done in an attempt to maintain tension on the cable (which stretches) as constant as possible for depth correlation purposes. (The exception to this practice is in certain hostile environments in which the tool electronics might not survive the temperatures on the bottom for the amount of time it takes to lower the tool and then record measurements while pulling the tool up the hole. In this case, 'down log' measurements might actually be conducted on the way into the well and repeated on the way out if possible.) The drum will usually operate at between 300 and 1800 m/h depending on the tool requirements; a simple gamma ray tool can

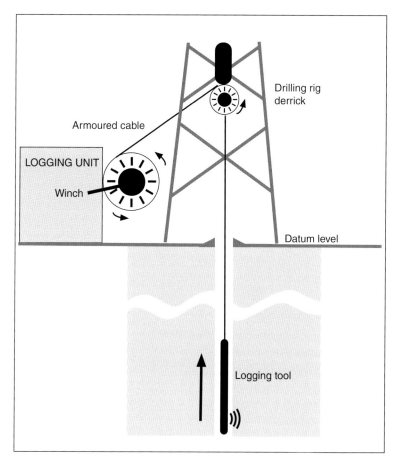

Figure 2.3 Schematic diagram of a typical set-up for running wireline logs. The logging unit, either a truck or Portakabin offshore, contains the surface control and data recording equipment. Set-up can take a few hours after the drillstring is retrieved.

be run at 1200 m/h whereas a density–neutron combination must be run more slowly. Most wireline measurements are recorded continuously even though the sonde is moving. Certain fluid sampling and pressure-measuring tools require that the sonde be stopped, increasing the chance that the sonde or the cable might become stuck. Under certain operational circumstances, such as when drilling an extended reach well, the logging tools may be pipe conveyed or pumped down to the zone of interest before commencing to log.

Logging tools (Table 2.1) all have different characteristics that may affect log quality, depending on the tool physics and the borehole environment: chief among these is the depth of investigation (Table 2.2). Most tools have only a shallow depth of investigation and may only read the flushed or

Table 2.1   Common wireline logging tools and acronyms.

Tool type/name	Physical measurement	Use
Environment		
Temperature (BHT)	Temperature	For resistivity calculations
Pressure (PRESS)	Fluid pressure	For formation volume calculations
Calliper (CAL)	Borehole diameter	Data quality, breakout
Lithology		
Gamma ray (GR)	Natural radioactivity	Shale indicator, correlation
Spectral gamma (NGS)	Component natural radioactivity	Depositional environment
Spontaneous potential (SP)	Electric potential	Permeable layers, R_w estimation
Porosity		
Sonic (BHC, LSS, DSI)	Acoustic velocity	Matrix porosity
Density (ΓDC, LDT)	Bulk density	Total porosity
Neutron (CNL)	Hydrogen index	Total porosity
Resistivity		
Induction (DIL, ILD)	Conductivity	Formation resistivity in oil-based mud, S_w
Laterolog (DLL)	Resistivity	Resistivity in water-based muds
Microlog (ML)	Resistivity of mudcake/ flushed zone	Indicator of permeability, thin bed resolution
Micro-laterolog (MLL)	Resistivity of flushed zone	Measures R_{xo}
Proximity log (PL)	Resistivity of flushed zone	Measures R_{xo}
Micro-spherically focused log (MSFL)	Resistivity of flushed zone	Measures R_{xo}
Other logs		
Formation tester (RFT, MDT)	Formation pressure and samples	Pressure measurements form the invaded zone and samples from the uninvaded zone
Sidewall coring tool (CST, RSWC)	Lithology	Percussion and rotary samples of borehole wall
Nuclear magnetic resonance (NMR)	Free fluid index	Porosity, moveable fluids, permeability
Borehole image logs	High-resolution resistivity and sonic	Structural and sedimentological

transition zone, especially if invasion has been significant. Resistivity tools have a large range of investigation depths because of the design, especially the focused tools; the deep penetration tools can investigate several feet into the formation.

The vertical resolution of tools can vary from a few centimetres to a couple of metres; high-resolution imaging tools can distinguish individual lamina in sandstones and shales. The vertical resolution is usually a function of the

Table 2.2 Logging tool depth of investigation and vertical resolution.

Tool type/TLA	Vertical resolution	Depth of investigation	Tool limitations
Spontaneous potential (SP)	2–3 m	N/A	OBM; $R_{mf} < R_w$
Gamma ray (GR)	60 cm	30 cm	Borehole rugosity
Spectral gamma ray (SGR)	100 cm	40 cm	Borehole rugosity
Photoelectric effect (PEF)	60 cm	5 cm	Barite mud
Dual laterolog (DLL) – D/M	60 cm/60 cm	1.5 m/45 cm	OBM; R_{mf}/R_w <2.5
Micro log (MSFL/ML)	5–10 cm	3–10 cm	OBM
Dual induction	<3 m/<2 m/<1 m	1.5 m/75 cm/40 cm	Res >200 Ω m;
DIL – D/M/S			R_{mf}/R_w <2.5
Array induction– D/M/S	1 m/1 m/60 cm	1.8 m/1.5 m/40 cm	Res >200 Ω m;
			R_{mf}/R_w <2.5
Bulk density (DENS)	45 cm	20 cm	Hole rugosity
Neutron porosity (CNL)	60 cm	25 cm	Stand-off
Sonic (BHC, AC)	60 cm	15 cm	Cycle skipping
Array sonic	1.5 m	15 cm	$V_{shear} > V_{mud}$
Dipole sonic	1.5 m	30 cm	–

Table 2.3 Standard wireline logs scales, units and ranges.

Log (track)	Measurement (units)	Left	Right
GR (1)	API	0	150
SP (1)	Millivolts (mV)	−10	+10
CAL (1)	Inches (in)	6	16
BIT_SIZE (BS) (1)	Inches (in)	6	16
RES (2)	Resistivity – logarithmic scale (Ω m)	0.2	200
SONIC (3)	Slowness (μs/ft)	140	40
DENS (2)	Bulk density (g/cm³)	1.95	2.95
NEUT (2)	Limestone porosity units (p.u.)	0.45	−0.15
PEF (2)	Barnes/electron (B/e)	0	10

distance between the source and detector; the shorter the distance, the greater is the resolution. The speed at which a logging run is made can also influence the quality of the results, through statistical fluctuations in either the measurement or sampling rates. Each tool has a specific set of criteria and conditions at which they function optimally.

Most service companies have a standard log presentation format for each tool combination; however, as the results are usually presented as digital records, the format is completely user driven. The main tool combinations are GR/SONIC, GR/RES and GR/DEN/NEUT (Table 2.3). In each case, the GR/SP/CAL/BIT_SIZE are in Track 1 on the left and the other combinations in Tracks 2 or 3 as appropriate. Scales, ranges and units for each log are also standard in most displays.

Table 2.4   Logging tool applications and limitations.

Logging tool	Applications	Interpretation limitations
Calliper	Borehole diameter; breakout	Check repeatability
Spontaneous potential (SP)	Correlation; R_w estimation	Shale effects, baseline drift
Gamma ray (GR)	Correlation; lithology; V_{sh} estimation	Radioactive sands effect
Spectral gamma ray (SGR)	Correlation; lithology; V_{sh} estimation	Compare U, Th, K ratios
Photoelectric effect (PEF)	Lithology and correlation	–
Dual laterolog (DLL) – D/M	Estimate R_t, R_{xo}, D_i in relatively salty mud	Clay-bound water; shaly sand interpretation
Micro log (MSFL/ML)	Permeability and moved hydrocarbon indicator; estimate R_{xn}	S_{xo} can be affected by invasion flushing
Dual induction (DIL) – D/M/S Array induction– D/M/S	Estimate R_t, R_{xo}, D_i in relatively fresh mud and OBM; reduced shoulder bed	Clay-bound water effect on S_w three readings will stack in impermeable beds
Bulk density (DENS)	Estimate porosity, lithology	Matrix and fluid density required for porosity
Neutron porosity (CNL)	Estimate porosity; indicate gas	Gas effect; check matrix lithology
Sonic (BHC, AC)	Measure compressional velocity and porosity	Matrix and fluid density required for porosity
Array sonic	Measure compressional and shear velocity	–

Table 2.4 presents the common logging tools and their primary applications and limitations; these will be discussed in more detail in following chapters. There are many other wireline tools available that provide specific information for reservoir evaluation, such as pressure measurements, borehole conditions and high-resolution images; their role will also be touched upon during the course of this book.

2.4   Well test data

In the past, a well test was undertaken to establish basic production information such as the nature of fluids, deliverability, reservoir pressures and permeability; however, the advent of precision gauges and PC-based analytical software has made the well test a crucial part of reservoir evaluation. When high-quality pressure measurements are available, the well test engineer is able to identify large-scale reservoir heterogeneities such as possible faults or facies boundaries; all interpretations are non-unique, however, and supporting evidence from seismic or geological data is required to provide a probable solution.

Of particular interest to the petrophysicist is the estimation of permeability from a well test; the rate of flow from a given producing interval provides the most illustrative value for use in dynamic simulation as it is an *in situ* measurement and represents more than a single or averaged value from core or log data. The calculated permeability–height value can be readily compared between wells to determine the most productive reservoirs and intervals.

2.5 Borehole environment

The borehole is a hostile environment for logging tools; high pressures and temperatures, aggressive chemical and physical conditions, due to the drilling fluids, formation fluids and the drilling process, geometry and condition of the hole all impact tool performance. The more sophisticated the tool, the greater are the physical limitations imposed by the electronics; tools have limitations based on the temperature and pressure ratings given by the equipment manufacturers.

Overburden, or lithostatic pressure, is the force due to the weight of the rocks above a reservoir zone and so is a function of the density of those overlying formations. Fluids in the pore spaces are also under a similar pressure; hydrostatic pressure is the normal fluid pressure where there is a direct connection to the surface. Hydrostatic pressure, also a function of fluid density, is generally one-third of the lithostatic pressure in normal situations (Figure 2.4). Where the fluids are not in communication with the surface, the normal situation in hydrocarbon reservoirs, an overpressure develops. The effective pressure on a reservoir is the difference between the overburden and formation pressure; if a reservoir is produced or a seal is breached, the pressure will eventually equilibrate. Reservoirs can also be underpressured, usually due to production depletion. It may be assumed that all pressure acting is isotropic, although this is not in fact the case owing to regional stress fields.

Drilling fluids (mud) are used to cool and lubricate the drill bit, remove drilled cuttings, counteract the fluid pressure in the rock and stabilize the borehole wall by building up a mudcake. Drilling mud is a suspension of clays, minerals and chemicals in a water- or oil-based medium, with specific elements to control the density, viscosity and gel strength – properties that maintain the efficiency of the fluid. The density of the drilling mud will vary depending on the expected and actual fluid pressures experienced while drilling, but there is always a built-in safety factor resulting in wells being drilled overbalanced. Because the drilling mud is at a greater pressure than the fluid pressure in the formation, mud flows into permeable units; this is called invasion. Invasion alters the rocks and fluids in the immediate vicinity of the borehole; drilling mud displaces the reservoir fluid with mud filtrate and leaves a mudcake on the borehole wall.

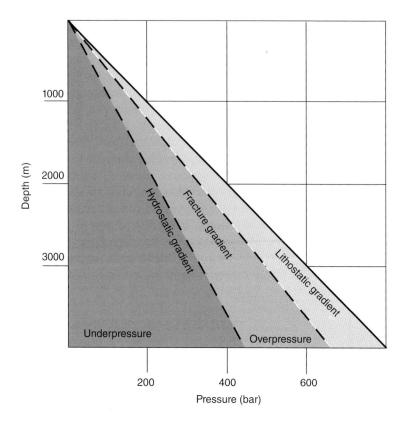

Calculations and conversions:

Pressure (psi) = MWppg*0.0519*TVDft

Pressure (bar) = MWg/cc*0.0981*TVDm

Figure 2.4   The pressure gradients that are to be expected in the subsurface as a well is drilled. The formation pressure lies generally between the hydrostatic and lithostatic gradients.

These changes to the initial reservoir conditions must be estimated and are used in the interpretation of the log measurements. The key environmental parameters are borehole diameter, drilling mud resistivity (R_m), extent and profile of the invaded zone and the resistivity of the flushed zone (R_{xo}) and uninvaded zone (R_t) (Figure 2.5). The longer a permeable unit is exposed to the drilling fluid, the greater is the invasion and the more complex the invasion profile as fluid segregation can take place (Figure 2.6).

The resistivity of the borehole and reservoir fluids varies with temperature; hence it is necessary to record the mean surface temperature and bottom hole temperature (BHT) of a well to calculate a temperature gradient against which the measurements can be calibrated. The geothermal gradient is a

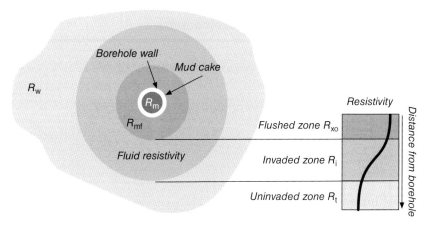

Figure 2.5 A representation of the zones of invasion around a vertical borehole and the resulting resistivity profile.

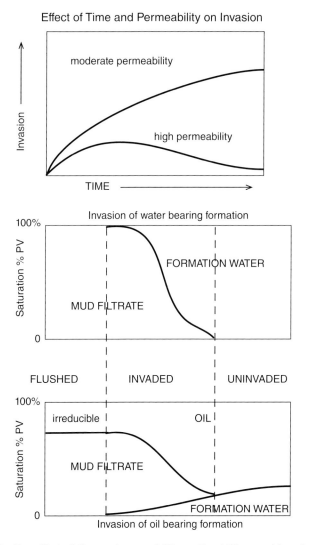

Figure 2.6 The effect of time and permeability on the drilling mud invasion profile.

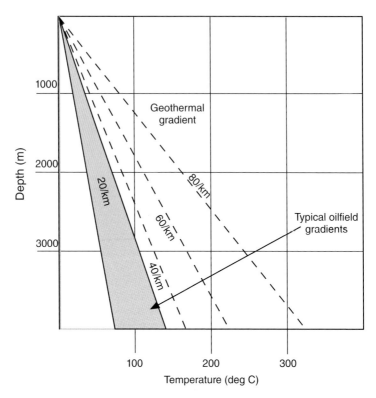

Figure 2.7 Different geothermal gradients showing increasing temperature with depth, with the zone of typical oilfield temperatures indicated.

measure the rate of increase in temperature with depth and can depend on the thermal conductivity of the rocks and the degree of tectonic/volcanic activity in a basin; a typical gradient is 1 °F per 70 ft (25 °C/km) (Figure 2.7). Shales, a thermal insulator, have a large gradient, whereas salts conduct heat efficiently and have a smaller gradient.

Sensors are attached to all logging tool strings to measure the maximum temperature in the well at the end of each well section and at total depth. The BHT is needed primarily to calibrate resistivity measurements, but is also used to detect flow into the well and for geochemical modelling. The actual temperature measured is that of the drilling fluid, not the formation. It can take many hours or days without mud circulation for the fluid to equilibrate to the formation. Hence the temperature measured by LWD tools, taken when mud is circulating, is not representative and should not be used in interpretation unless calibrated. Even wireline measurements usually require some temperature correction before becoming usable; this is normally done using a Horner plot, which aggregates the temperature from several logging runs against the time since circulation stopped.

Borehole geometry – size and shape – is measured using a variety of different calliper tools, including those with simple mechanical arms, more accurate four-arm tools that capture orientation and acoustic measurements that allow the borehole wall to be displayed in 3D. The mechanical tools have 2–6 articulated arms that send a resistivity signal from a sensor pad at each end that is in contact with the borehole wall. Acoustic callipers record an ultrasound signal from the borehole wall as it rotates rapidly, sending and receiving a signal from the same transducer. In this remote sensing tool, accurate knowledge of the velocity of sound in the drilling fluid is required. The measurements made by any such tools provide information on washouts, borehole breakout due to local or regional stress fields and also the presence of fractures.

2.6 Summary

The main reason to drill a borehole is to acquire data. In the simplest form, this may just be whether hydrocarbons, water or even mineral concentrations are present. However, during the drilling of the borehole, there is plenty of basic information to be gained by using simple observations or sophisticated measurements while drilling. The role of the mud logger is often underestimated and the importance of the mud log as a daily record of all the events during the drilling of the well is often ignored, but for the petrophysicist it should be the first source of reference data. The acquisition of core data is always divisive; everyone knows the value of a core, but justifying the cost in terms of time and money is sometimes difficult. Some companies will never core an exploration well because if successful they will expect to return for appraisal, but this is often a false economy, because the calibration of all the other information acquired is lacking. Wireline logging programmes take on a life of their own when devised as part of a decision tree: different combinations of logs depending on the preliminary outcomes of the well. Basic log suites should be consistent for any exploration and appraisal programme so that all wells may be compared equally; in the development phase, the emphasis correctly shifts to low-cost acquisition until the unexpected happens and additional data are required! Through all the phases, the acquisition of dynamic data by pressure measurements or well tests should be an essential requirement.

3

Rock and Fluid Properties

In a reservoir, fluids flow through an interconnected network of pores created by the distribution of grains of quartz, feldspar, lithoclasts or carbonate grains and fragments; the fluids, water, oil or gas, flow during hydrocarbon migration and production. Microscopic forces (gravity, viscosity and capillarity) are effective during the initial distribution of the fluids and so the study of the pore system at this scale is essential to understanding the rock properties of the reservoir and fluids. It is at this point that the representative scale of the data becomes important in a petrophysical study: integration of data from the microscopic, macroscopic and megascopic scale is a major challenge. When we analyse a rock sample or interpret a continuous log, we are only sampling a miniscule volume of the reservoir and often try to extrapolate these data over a much larger volume. The concept of *representative elementary volume* (REV) was introduced to try to capture the many scales of data and how they may be used in reservoir characterization (Figure 3.1). The REV is the smallest volume over which a measurement can be made that represents the whole volume of a sample (Bear, 1972). Scales of measurement become an important aspect of reservoir characterization, whether you start with petrographic data from a thin section or wireline log data, integration of the data and the avoidance of sample bias is a fundamental prerequisite.

3.1   Controls on rock properties

Looking at clastic systems first, grain sorting and packing define the pore network more than the grain size alone. Perfectly rounded, equally sized spheres packed in a cubic pattern, independent of grain size, have a porosity of 47.6%, whereas the same spheres packed in an orthorhombic pattern have a porosity of 26.0% (Figure 3.2). By mixing the grain sizes, such that the larger pore spaces become filled with smaller grains, the porosity will be reduced.

Petrophysics: A Practical Guide, First Edition. Steve Cannon.
© 2016 John Wiley & Sons, Ltd. Published 2016 by John Wiley & Sons, Ltd.

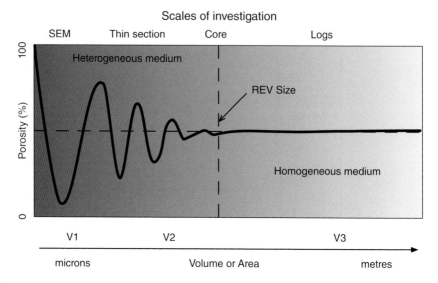

Figure 3.1   The representative elementary volume (REV) and scales of investigation and measurement in heterogeneous and homogeneous media. Source: Bear (1972). Reproduced by permission of the American Elsevier Publishing Company.

The degree of sorting is a function of the depositional mechanism, energy and the source of the sediments; moderately or well-sorted grains often reflect a period of prior sorting before deposition or a second depositional cycle. The pore network of a clastic reservoir is generally intergranular in nature; however, compaction and diagenesis can alter the porosity significantly, and seldom for the better.

Carbonate rocks are often grain dominated, but undergo extensive post-depositional chemical changes that affect the primary pore network. The grains may be organic (bioclasts) or inorganic (pellets, ooids) providing both inter- and intragranular pore systems. Calcium carbonate in the form of aragonite or calcite is the primary constituent of most marine organisms that form limestone reservoirs; aragonite is chemically less stable and is quickly altered to calcite. Dolomite is a magnesium carbonate and is generally associated with altered limestone, changing the crystal structure and producing enlarged pore systems; these may not be as well connected as the original system. The complex geological, physical and chemical nature of carbonate rocks often results in substantial pore systems that are ineffective as producing reservoirs without some type of stimulation to connect the matrix, where the bulk of the hydro-carbons are stored, with the primary flow paths.

A detailed petrographic study, including thin section analysis, X-ray diffraction (XRD) and scanning electron microscopy (SEM) can greatly improve the petrophysical description of a reservoir (Figure 3.3). Thin section

(a)

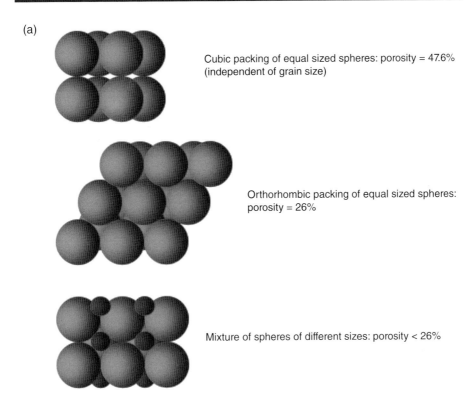

Cubic packing of equal sized spheres: porosity = 47.6%
(independent of grain size)

Orthorhombic packing of equal sized spheres:
porosity = 26%

Mixture of spheres of different sizes: porosity < 26%

(b)

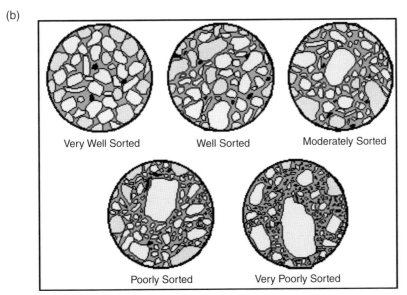

Figure 3.2 (a) The impact of grain size, sorting and packing on porosity in typical clastic rocks. (b) Typical visual estimation of degrees of sorting in sandstones. (For a colour version see Plates section)

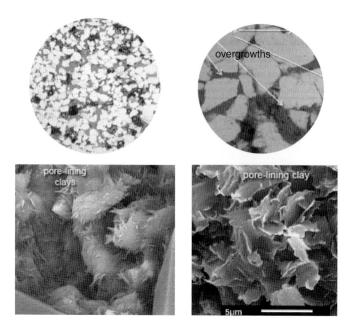

Figure 3.3   Examples of petrographic thin sections and SEM images: porosity is stained blue. (For a colour version see Plates section)

microscopy provides the basic investigative method for characterizing both the mineralogy of a reservoir rock and the textural properties: grain size, sorting and sphericity. A detailed description of the sample can reveal information about the diagenetic history of the rock and the main porosity types; thin sections are usually stained to aid a visual estimate of porosity. The degree of lithification of the rock can also be estimated by studying grain contacts and the degree of compaction.

The use of SEM allows a deeper investigation and characterization of a pore system in terms of the shape, size and geometry and mineralogy of the constituent elements. A carefully sampled set of representative data will at least qualitatively help characterize a reservoir for a petrophysical study. SEM is used in the recognition and distribution of different clay types, all of which can affect the quantitative interpretation of wireline data. The presence of certain clay types, such as illite, may also affect reservoir performance by reducing the permeability of a reservoir. The use of XRD is equally important in recognizing the elemental constituents of a reservoir and for understanding their influence on wireline logs. The presence of heavy minerals such as pyrite or siderite will strongly influence the density and, in the case of pyrite, the resistivity readings, ultimately impacting on the calculation of water saturation if not recognized.

3.2 Lithology

The two main logs available for lithology determination are the *spontaneous potential* (SP) and the *gamma ray* (GR) or *natural gamma tool* (NGT). Other measurements that may be influential in determining the lithology are the acoustic response, bulk density and photoelectric absorption. The tools that make these measurements are run primarily to determine porosity and will be described later.

3.2.1 Spontaneous potential

The SP is one of the earliest wireline tools and was developed in the 1940s. The tool measures the direct current voltage that develops naturally between a moveable electrode in the well and a second, fixed electrode at the surface (Doll, 1948). The units of measurement are millivolts (mV). The tool response is created by electrochemical variations between the borehole fluid and reservoir rock brought about by salinity differences between the mud filtrate and formation water within permeable beds. The response can be positive (to the right) or negative (to the left), depending primarily on the relative salinity of the mud and formation. If the salinity of the mud filtrate is less than the formation salinity, the deflection is negative – this is the normal situation. Where there is little difference in salinity, the deflection from the base line is minimal. Be aware that over a long borehole section the shale baseline can drift, either positively or negatively, and will need to be corrected. Obviously if the mud is not conductive, i.e. oil-based mud, no measurement can be made.

The SP currents that flow between the two fluids are caused by an electromotive force (EMF) comprising both electrochemical and electrokinetic components. The former is a function of the positive sodium ions (Na^+) that are able to flow through shale from the more concentrated NaCl solution to the weaker one, whereas the negative chloride ions (Cl^-) are blocked, creating an electrical current and the potential difference. The potential across the shale is termed the membrane potential. In clean sands, the total electrochemical EMF (E_c) has the following relationship:

$$E_c = -K \log \frac{a_w}{a_{mf}} = -K \log \frac{R_w}{R_{mf}}$$

where a_w and a_{mf} are the chemical activities of the formation water and mud filtrate, respectively at formation temperature and K is a coefficient proportional to the absolute temperature equivalent to 71 at 25 °C. R_w and R_{mf} are the resistivity of the formation water and mud filtrate, respectively, thus allowing the SP to be used quantitatively to estimate formation water salinity.

The electrokinetic component of the SP is produced when an electrolyte flows through a permeable non-metallic porous medium, in this case the mud-cake that develops opposite permeable formations. The size of this potential is dependent primarily on the pressure difference producing the flow and the resistivity of the electrolyte. The scale of these effects is negligible in most cases, but can become apparent where a depleted reservoir zone is penetrated or drilling fluids with a high mud weight are used to manage overpressured sequence.

The SP is used to detect permeable beds and their boundaries, determine formation water resistivity and estimate the volume of clay in permeable beds. The recorded value is influenced by bed thickness, bed resistivity, shale content, hydrocarbon content, borehole diameter, drilling mud invasion and R_{mf}/R_w. The SP is often used to determine the sand from shale in clastic sequences and also the volume of clay in permeable sand beds; in water-bearing sands, the degree of SP reduction is related to the amount of shale in the formation (Figure 3.4). The response of the SP in a shale sequence is generally constant, defining a shale baseline against which positive $(R_{mf} < R_w)$ and negative $(R_{mf} > R_w)$ departures are recorded that represent permeable layers. As the tool is responding to the difference between the resistivity of the mud filtrate and the formation fluid, the magnitude of the deflection is related to the ratio of the fluid resistivity, not the permeability. The SP is also the only tool to identify categorically fresh water zones in a borehole; resistivity tools remain ambiguous in this observation.

The static spontaneous potential (SSP) is a measure of the SP corrected for bed thickness and clay content in sands; the value can be calculated or read off a chart. SSP is used in the estimation of formation water resistivity; this should only be done for known water-bearing sands in which the SP and SSP are the same. The PSP (pseudostatic spontaneous potential) is a measure of the maximum SP in a shaly formation and can be used with SSP to estimate the volume of shale (V_{sh}) in a permeable zone using the following equation:

$$V_{sh} = \frac{PSP - SSP}{SP_{sh} - SSP}$$

The SP response is suppressed in the presence of hydrocarbons; however, this phenomenon cannot be used to quantify the hydrocarbon saturation.

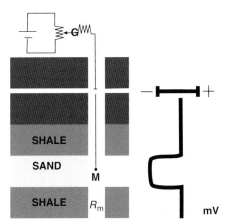

Operating principle - measures electric potentials that occur naturally in the borehole due to mud filtrate invasion into the reservoir (electro kinetic) and due to differences in ion concentration between the mud filtrate and the formation water (electro chemical).

Principal Application - Correlation, detection of permeable beds, determination of R_w, shaliness and bed thickness.

Limitations - Loses detail when R_w approaches R_{mf}; loses character if R_t/R_m increases; deflection subdued as shaliness increases or bed thickness decreases; not suitable in oil based mud or large diameter boreholes.

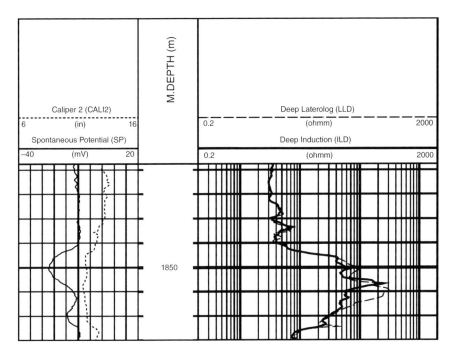

Figure 3.4 The SP (spontaneous potential) log, the simplest of all electrical measurements made in a borehole.

One further note of caution: the quality of the SP is dependent on the quality of the fixed electrode or earth. Onshore this is a simple matter of plunging the electrode into the ground; offshore it is not a trivial issue and the recorded log is often of little value to the interpreter.

3.2.2   Gamma ray

A gamma ray tool measures the natural radioactivity in a formation, responding to the presence of uranium-, potassium- and thorium-rich minerals. Quartzose sandstone with low concentrations of these minerals will record a minimal response, but when detrital clay is present or there is a high proportion of potassium feldspar, glauconite, heavy minerals or mica, there is a stronger response. As the clay content increases, the gamma ray response increases; organic-rich marine shale commonly has the greatest response as it contains significant amounts of uranium-rich minerals generated by the reduction of decaying organic matter. Carbonates rocks generally have a low gamma ray response unless they contain significant detrital clay or uranium.

Nearly all gamma radiation emitted naturally is from the radioactive potassium isotope of atomic weight 40 (^{40}K) and by the radioactive elements uranium and thorium. The number and energies of each element are distinctive and can be used to discriminate between them. This fact is used in the spectral gamma ray tool that uses selective energy windows to deconstruct the total gamma response into these separate elements. Commonly the results are displayed as a total gamma ray (SGR) and a computed gamma ray (CGR), which is total gamma minus uranium. Either type of gamma ray log may be used for estimating the volume of detrital clay in a reservoir, but if there is known to be a high uranium content this should be discounted when performing the exercise, and likewise if the reservoir is an arkosic sandstone with a significant potassium feldspar content, highly micaceous or contains volcanic ash.

The GR sonde contains a detector, usually a scintillation counter, typically a crystal of sodium iodide, which measures the gamma radiation emitted close to the borehole wall (Figure 3.5). Because of the relatively small size of the counters, good resolution of formation variation is achieved and it is normally run in all tool strings as an aid to correlation. The primary calibration standard for GR tools is the American Petroleum Institute (API) test facility in Houston and logs are normally presented as API units. Older tools used a scale of micrograms of radium-equivalent per ton of formation; former Soviet Union tools continued to use this scale until the late 1990s and conversions are available for these older tools. Tools are also field calibrated to the API standard before being sent out on a job. Although largely insensitive to logging speed, the simple gamma ray tool records more 'counts per second' at slower speeds, improving the overall accuracy of the measurement; the spectral gamma ray is very sensitive to logging speed, however, and is normally run with density and neutron combinations. The tool response requires correction for borehole size and rugosity and the density and make-up of the drilling fluid, as these can impact on the capture of gamma rays from the formation, especially in washed out intervals or if the mud is particularly heavy.

STANDARD GAMMA RAY

Operating principle - measures the natural radioactivity of the formation.

Principal applications - correlation; lithology identification; shale content; sand count, depth control.

Limitations - log quality depends on time constant and instrument sensitivity; potassium bearing mud will give false readings.
Corrections required for borehole size, mudweight and salinity in older tools

SPECTRAL GAMMA RAY (NGT)

Operating principle - to distinguish between the three families of naturally radioactive elements (k,Th,U) and to assess their respective proportions. The counting is done selectively in narrow energy windows.

Principal applications - determination of depositional environments; identification of organic-rich shales, source rocks.
Uranium free computed log should be less than or equal to the standard gamma ray

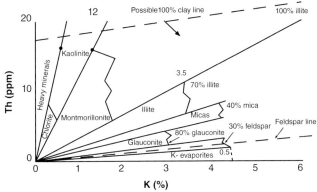

Graph showing the distribution of clay minerals, heavy minerals and evaporites in terms of potassium and thorium content (redrawn from Quirein et al, 1982)

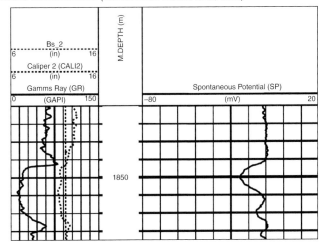

Figure 3.5 Gamma ray logging measurements of both the normal and spectral gamma tools.

The presence of radioactive potassium chloride, KCl, as a mud additive can also affect the gamma ray response, especially where the chemical has invaded permeable intervals or has built up in the mudcake.

The GR is used primarily to define the volume of shale (V_{sh}) in a sequence, especially where the SP response is distorted or if oil-based mud is being used. Shale is composed of clays, silt and mud that is lithified during compaction due to burial and is frequently laminated, forming discrete structural layers. Shale comprises 50–70% clay minerals such as kaolinite, smectite and illite, with the remaining particles being predominantly silt-grade quartz grains. A distinction should be made between shale volume and clay volume, terms that are often used indiscriminately: shale refers to a rock volume and clay to a specific component; this becomes more significant when dealing with clay-rich sandstone reservoirs.

When comparing GR logs from a number of wells, they should all be normalized to a common scale, as each tool will have been calibrated individually. This is especially true if being used for correlation or quantitative calculations. Calculation of the gamma ray index (GRI) is the first step in the process of estimating shale volume where there is a linear relationship:

$$GRI = \frac{GR_{sh} - GR_{min}}{GR_{log} - GR_{min}} = V_{sh}$$

As the relationship can vary geographically and stratigraphically, a number of non-linear relationships for younger rocks (Larionov, Steiber and Clavier) have been developed over the years, but their application is limited today, unless a specific need is identified.

Typical GR responses seen in commonly occurring rocks and minerals are given in Table 3.1. The availability of GR logs in most wells has led to their being used extensively for correlation and for sedimentary sequence analysis. The inference is that the GR response is associated with variable proportions of shale or clay minerals in sandstone: often the response is in fact a function of the depositional energy of the sediments, clean sands reflecting high-energy conditions in which clays are removed and then deposited preferentially in lower energy environments. This relationship is apparent in fluvial channels and their overbank deposits and in shoreface sands and their offshore equivalents (Figure 3.6). The resulting GR and SP patterns have been well documented since the 1950s, when they were first categorized by Shell Research for the Gulf of Mexico sedimentary sequences. Modern 'sequence stratigraphers' use similar techniques today to characterize depositional systems in a context of fluctuating sea levels (*plus ça change!*).

Table 3.1 Typical GR responses seen in commonly occurring rocks and minerals.

Type	Mineral	Composition	API units
Pure mineral	Calcite	$CaCO_3$	0
	Dolomite	$CaMgCO_3$	0
	Quartz	SiO_2	0
Common lithologies	Limestone		5–10
	Dolomite		10–20
	Sandstone		10–30
	Shale		80–140
	Organic-rich shale		150–200+
Evaporites	Halite	NaCl	0
	Anhydrite	$CaSO_4$	0
	Gypsum	$CaSO_4.H_2O$	0
	Sylvite	KCl	500
	Carnalite	$KCl, MgCl_2.6H_2O$	200
	Langbeinite	$K_2SO_4, 2MgSO_4$	275
	Polyhalite	$K_2SO_4, MgSO_4, 2CaSO_4.2H_2O$	180
	Kainite	$MgSO_4, KCl.3H_2O$	225
Others	Sulfur	S	0
	Lignite	C	0
	Anthracite	C	0
	Mica, feldspar		100–170

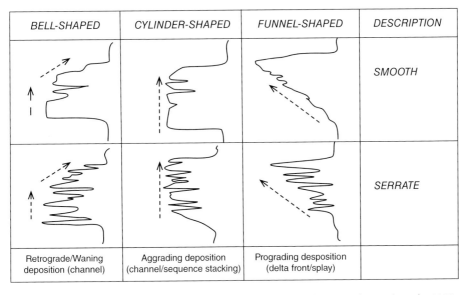

BELL-SHAPED	CYLINDER-SHAPED	FUNNEL-SHAPED	DESCRIPTION
			SMOOTH
			SERRATE
Retrograde/Waning deposition (channel)	Aggrading deposition (channel/sequence stacking)	Prograding desposition (delta front/splay)	

Figure 3.6 Typical gamma ray or SP log profiles and the descriptive terms in use since the 1950s to infer the environment of deposition.

3.3   Porosity

Porosity (Phi or ϕ) is defined as the ratio of the volume of pore (void) space (V_p) to the total volume of rock (V_t). It is a dimensionless property usually expressed by geologists as a percentage and engineers as a decimal fraction; we shall use the latter expression, unless specified. Even such an apparently simple property can be a difficult one to quantify when taking into account the mineralogy of the rock and also its diagenetic history.

First, porosity may be described as primary or secondary depending on whether mineral dissolution has occurred during lithification; this may enhance total porosity but may not result in improved flow properties as such pores are often isolated. Secondary porosity is usually of greater importance in carbonate reservoirs because of the higher solubility of calcareous minerals. A 'fractured' reservoir may also show a small porosity enhancement or the fractures may be the only source of fluid storage volume; fracture porosity may be difficult to identify and quantify without additional information from cores, image logs or seismic data.

Second, porosity can be reported as either total (PhiT) or effective (PhiE), depending on the source of the measurement; core data are generally assumed to be 'total' because of the cleaning and drying process in the laboratory, but a log-derived measurement could be either effective or total depending on how it has been derived. Some logs also measure shale porosity, and this must also be considered when determining the effective porosity of a reservoir (Figure 3.7).

3.3.1   Core porosity

Porosity measurements on core material usually rely on accurately estimating the pore volume of the sample using a gas expansion method. The result will be somewhere between effective and total porosity measured by a wireline log depending on the cleaning and drying process used on the sample. It is important, therefore, to identify the source of the data before they are used in making an interpretation.

Core samples provide an accurate and repeatable measurement of porosity: a cleaned plug is weighed and measured to calculate the bulk volume and grain density prior to insertion into a helium porosimeter. Using Boyle's law, the porosimeter measures the connected porosity of the sample or at least as much of the pore space that the expanding helium gas can occupy; microporosity or isolated pores may not be fully saturated. The value of the grain density can give an indication of the quality of the result; a sample with a lower than expected density for the given lithology may still contain water or hydrocarbons in the smallest pores, thus resulting in a lower porosity measurement. Generally this is a quick and reliable measurement of porosity

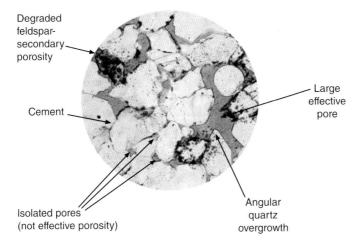

LITHOLOGY	POROSITY RANGE (%)
Unconsolidated sands	35–45
Reservoir sandstones	15–30
Cemented sandstones	5–15
Conglomerate	1–15
Shales/Clays	0–45
Non-vuggy limestones	5–10
Vuggy limestones	10–40
Dolomite	10–30
Chalk	5–40
Granite	<1 (matrix)
Basalt	0–5 (vesicular)

Figure 3.7 Types of porosity seen in thin section and typical porosity ranges found in different potential reservoirs.

at ambient conditions; an experimental rig can be used to replicate reservoir pressure to obtain an overburden-corrected result.

Core porosity is usually taken as the definitive measurement of a reservoir's porosity against which log estimations are calibrated. However, there are a number of issues that should be considered when evaluating the accuracy and representativeness of the results.

3.3.1.1 Accuracy
When evaluating data from a number of wells, it is to be expected that different laboratories or different techniques may have been adopted; this may result in variability in data from the same reservoirs that is not predictable. It is always worthwhile reviewing data in terms of the acquisition parameters where possible.

3.3.1.2   Support volume

Laboratory measurements are made on an infinitesimally small volume of reservoir that may not be as representative as we might wish; issues with representative support volumes of data abound within reservoir characterization studies. This is especially true for very heterogeneous reservoirs, where systematic variations in reservoir quality become less and less predictable. Typically, carbonate reservoirs are more subject to these issues than clastic reservoirs; however, coarse-grained and conglomeratic reservoir can also show such phenomena.

3.3.1.3   Overburden correction

In general, well-indurated samples require little or no correction; however, younger or poorly consolidated reservoirs may require significant correction. These types of reservoir will often require special treatment throughout the core handling and analysis procedure. Empirical compaction corrections exist that can be used to correct ambient core measurements; or the porosity of representative samples can be measured at reservoir conditions to establish a reservoir-specific correction. The application of overburden corrections is dealt with later in this chapter.

3.3.2   Log porosity

There are a number of logs that 'measure' porosity, although none actually does this directly. The three commonest tools are the sonic log, which measures the acoustic response of a formation, and the density and neutron logs, which make nuclear measurements. When these tool responses are combined, two or three at a time, lithology can also be determined, along with a representative porosity interpretation. A fourth tool, NMR, measures the magnetic resonance of hydrogen nuclei in the reservoir fluids. Pulsed neutron logs are able to determine porosity from behind casing and may be used after a period of hydrocarbon production as part of a surveillance programme.

3.3.2.1   Sonic

The sonic log measures the interval transit time of a compressional sound wave travelling through the formation along the axis of the borehole wall. A compressional sound wave can travel in solids, liquids and gases; however, the fastest path for the wave to follow is through the solid (Table 3.2). As a result, the sonic tool records the matrix porosity of the formation; therefore, in vuggy rocks, such as many carbonates and sandstones that have a component of secondary porosity, other methods must be used to estimate the total porosity of the formation.

Table 3.2 Sonic velocities and interval transit times for different matrix types.

Lithology/fluid	Matrix or fluid velocity (ft/s)	Δt_{matrix} or Δt_{fluid} (µs/ft)
Sandstone	18,000–19,500	55.5–51.0
Limestone	21,000–23,000	47.6
Dolomite	23,000–26,000	43.5
Anhydrite	20,000	50.0
Salt	15,000	66.7
Casing (steel)	17,500	57.0
Freshwater mud filtrate	5,280	189
Saltwater mud filtrate	5,980	185

The sonic tool comprises one or more ultrasonic transmitters and two or more receivers positioned vertically to minimize and compensate for the effects of borehole rugosity (Figure 3.8). The interval transit time (Δt) is the reciprocal of the velocity of the sound wave passing through the formation and is measured in microseconds per foot or metre (µs/ft or µs/m). Δt is dependent on lithology and the fluid-filled porosity of the formation. The sonic log is usually presented in track 3 of a standard API display at a scale of 40–140 µs/ft (slowness increasing to the left of the display).

Modern sonic scanning tools are able to provide a 3D acoustic characterization of a formation for both rock physics and geomechanical interpretation. Multiple monopole and dipole transmitters generate compressional, shear and Stoneley (tube) waveforms, which, after processing, provide accurate slowness values that provide information for well completion and borehole stability studies. These tools can be run in open or cased boreholes and can provide information on geomechanical anisotropy in the reservoir.

All sonic tools have technical and physical limitations due largely to the triggering of the initial sonic pulse and the distance travelled to the multiple receivers. Noise can be generated mechanically or caused by stray electrical interference and will be picked up by the receivers; if severe enough, the stray signal will mask the true response of the first acoustic cycle at the receiver. Subsequent signal cycles, although having larger amplitude, travel further through the formation and can be weaker at the receiver resulting in a cycle 'skip' such that a higher Δt value is recorded, producing a 'spike' in the log. In larger boreholes, the shortest distance between the transmitter and receiver may be through the drilling fluid; in this case, the first cycle does not penetrate the formation.

Today, the main use of the range of sonic tools is in support of seismic and geophysical interpretation, first in making a well tie between time and depth measurements for depth conversion, and second for calculating interval velocities and also for acoustic impedance studies as part of predictive rock physics experiments. Further discussion can be found in Chapter 8.

Vertical resolution/DOI: 30cm/15cm
Logging speed: 1200 m/h
Can be run open hole in WBM or OBM
Tool should be eccentred in large diameter holes

Upper Transmitter   Sonic and density data are used to determine acoustic impedance and reflection coefficients at bed boundaries.

Used for fluid substitution models to predict presence of hydrocarbons and porosity.

Receivers

Compressional and shear sonic data used for geomechanical studies on rock strength and to determine abnormal pore pressures ahead of drilling.

Lower Transmitter

Specially adapted sonic tools are used for QC of cement bonds.

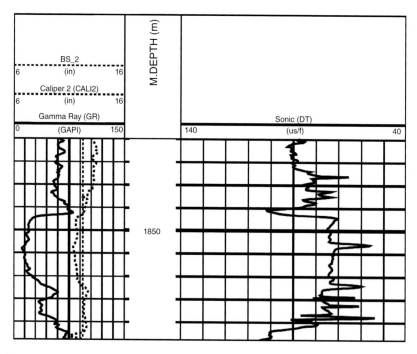

Figure 3.8   The borehole-compensated sonic tool: mode of operation, applications and typical display.

3.3.2.2 Density

The density log measures the bulk density of the formation; that is, the density of the rock plus the fluids contained in the pores. Density is measured in g/cm^3 and is by convention given the symbol ρ (rho). The density log is usually presented across tracks 2–3 of a standard API template along with the neutron and PEF logs: the scale is usually 1.95–2.95 g/cm^3. To calculate porosity from a density tool, it is necessary to know the matrix density and the density of any fluids in the pore space. The matrix or rock density is constant for a given pure lithology such as limestone or sandstone, in other words a solid with no porosity (Figure 3.9).

The density tool is skid-mounted to maximize contact with the borehole wall and consists of a radioactive source, such as cobalt (^{60}Co) or caesium (^{137}Ce), which emits medium-energy gamma rays, or in more modern tools an accelerator source; tools are built with two detectors, about 50 cm from the source, to compensate for borehole rugosity. The emitted gamma rays collide with electrons in the formation and each collision results in a loss of energy from the gamma particle (Compton scattering). The scattered particles that return to the detectors in the density tool are measured in two ranges: a higher energy range affected by Compton scattering and a lower energy range governed by the photoelectric effect (PEF). The number of higher energy range particles returning to the detector is proportional to the electron density of the formation density through a constant (Tittman and Wahl, 1965). The porosity is derived from this relationship with bulk density. The combination density–PEF tool is referred to as the lithodensity (LDT) or sometimes the photodensity tool (MPD). The density tool has a relatively shallow depth of investigation (~35 cm) and as a result is held against the borehole wall (eccentred) to maximize the formation response. Also recorded is a density correction curve (Drho) that indicates the level of correction applied during processing to account for borehole rugosity; intervals with corrections above ~0.20 g/cm^3 should be ignored in subsequent porosity calculations. The density correction is displayed in track 3 of a standard API plot.

The response of the low-energy returning particles is governed by the formation (lithology) and is nearly independent of porosity. The response is measured in barns/electron (B/e) (Table 3.3). The PEF response is used to identify evaporite minerals, evaluate complex lithologies such as mixed clastic and carbonates, detect gas-bearing zones and determine hydrocarbon density. Drilling muds weighted with barite make the results unusable as barium has an atomic number of 56 and therefore a much higher electron density than most common minerals.

3.3.2.3 Neutron

Neutron logs measure the hydrogen concentration in a formation, the hydrogen index (HI); the commonest source of hydrogen in the formation

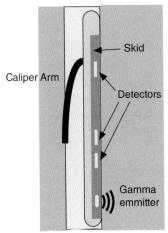

DENSITY

Vertical resolution/DOI: 100cm/5cm

Logging speed: 1200m/h; enhanced processing requires slower speeds

Can be run open hole in WBM or OBM

Main uses are for porosity and lithology identificaon combined with neutron log. Combined with sonic log to determine acoustic impedance

PEF

Vertical resolution/DOI: 100cm/1.5cm

Logging speed: 1200m/h; enhanced processing requires slower speeds

Used for mineral and lithology identification

Lithology/Fluid	ρ_{matrix} or ρ_{fluid} g/cm³	PEF (B/e)
Sandstone	2.644	1.81
Limestone	2.710	5.08
Dolomite	2.877	3.14
Gypsum	2.355	3.42
Anhydrite	2.960	5.05
Salt	2.040	4.65
Fresh water	1.0	0.36
Salt water	1.15	0.81
Oil	0.81	0.13
Barite	4.48	267

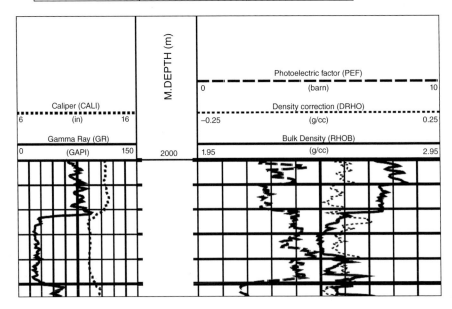

Figure 3.9   The formation density tool: mode of operation, applications and typical display.

Table 3.3 Matrix density and photoelectric effect values for common lithologies.

Lithology/fluid	ρ_{matrix} or ρ_{fluid} (g/cm³)	PEF (B/e)
Sandstone	2.644	1.81
Limestone	2.710	5.08
Dolomite	2.877	3.14
Gypsum	2.355	3.42
Anhydrite	2.960	5.05
Salt	2.040	4.65
Fresh water	1.0	0.36
Salt water	1.15	0.81
Oil	0.81	0.13
Barite	4.48	267

will be water or hydrocarbons (Figure 3.10). In shale-free rocks where the pore space is filled with water or oil, the neutron log directly measures liquid-filled porosity. Where the pores are filled with gas the concentration of hydrogen is reduced, resulting in a lower porosity reading from the tool, the so-called gas effect; there is a 'cross-over' with the density log when the neutron porosity is less than the bulk density in a porous and permeable zone. The neutron log is usually plotted across tracks 2–3 of a standard API display in conjunction with the density log; the display scale is normally 0.45 to –0.15 in limestone porosity units.

A chemical source in the tool, usually composed of americium and beryllium, continuously emits 'fast' neutrons that collide with the atomic nuclei in the formation. With each elastic collision, the neutron loses energy and eventually the neutron is absorbed by a nucleus and a gamma ray is emitted. The maximum energy loss occurs when a neutron collides with a hydrogen atom because they have similar atomic mass; thus the tool response is controlled by the formation hydrogen content, which can be directly related to the porosity for a given lithology. Neutron log responses vary depending on the type of source and the spacing between the source and detector: the effects of such variations are usually processed out, but any tool corrections should be made with full knowledge of the tool type and manufacturer. Neutron data are not measured in basic physical units, but in porosity units usually calibrated to a standard limestone or sandstone response exhibiting zero porosity. Where shale is present in the formation, the neutron log responds to the water trapped in the clay particles, resulting in an overestimation of formation porosity.

The hydrogen index (HI) is based on the number of hydrogen atoms per unit volume of rock divided by the number of hydrogen atoms per unit volume of pure water at surface conditions – a proxy measure of the porosity of a rock. If we have a tool measuring zero porosity in a pure limestone, the HI is zero because

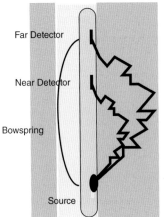

COMPENSATED NEUTRON LOG

Vertical resolution/DOI: 100cm/15cm

Logging speed:1200m/h; enhanced processing requires slower speeds

Can be run open hole in WBM or OBM

Acquisition software calculates neutron porosity referenced to a specific lithology, normally limestone. In the presence of gas the results are erroneous.

Requires borehole and environmental correction, especially pressure and temperature

Generally used in combination with sonic and/or density tools to estimate lithology and porosity. Neutron porosity is greater than actual porosity in shaly sandstones.

PULSED NEUTRON LOG

Uses a pulsed neutron generator and four partially shielded detectors that only count epithermal neutrons that have passed through the formation giving a more accurate measurement; epi-thermal neutrons have a higher energy count than thermal neutrons. Tool is not used much in open-hole interpretation but as cased hole log.

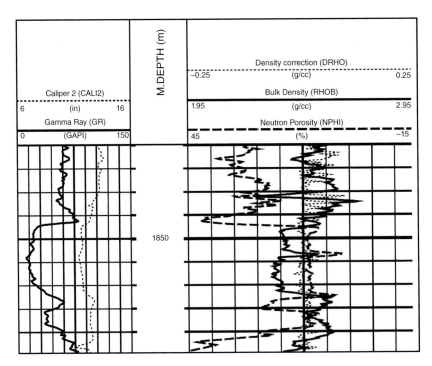

Figure 3.10 The compensated neutron porosity tool: mode of operation, applications and typical display.

there are no hydrogen atoms in the matrix; thus we have a fixed point where HI is zero and porosity is zero. Combined with the known value for unit volume of water where porosity is unity, we are able to scale our neutron response in a porous limestone filled with water. These results, however, are affected by both hydrocarbons and excess chlorine (salt) and therefore need to be corrected.

3.3.2.4 Nuclear magnetic resonance

Nuclear magnetic resonance (NMR) logging uses the magnetic moment of hydrogen atoms to determine directly porosity and pore size distribution as an estimate of permeability. The NMR tool responds exclusively to protons and the signal amplitude is directly proportional to the quantity of hydrogen nuclei present in the rocks as water or hydrocarbons, giving a value of porosity that is free from lithology effects; porosity is estimated from the rate of decay of the signal amplitude (Figure 3.11). Permeability can be derived from an empirical model linked to irreducible water saturation. The NMR display is completely different from that of other tools, encompassing raw and processed data to give a range of outputs including interpreted lithology, porosity, permeability and an estimation of volume of oil, gas and water.

Each service company has developed a different approach to the measurement using the same physical properties. In essence, a permanent magnet is used to align the proton spin axis of the hydrogen in a reservoir fluid, then a radio transmitter is used to disturb the spin axis and a receiver records the electromagnetic signal emitted as the protons precess back to the original spin axis: these are termed pulsed NMR tools. The emitted signals are observed either as parallel (longitudinal) or perpendicular (transverse) to the direction of the applied magnetic field and are expressed as time constants related to decay magnetism of the total system. The initial time constant, T_1, is called the longitudinal relaxation time and measures time taken for polarization (alignment) of the protons in the reservoir fluid. Once the magnet is turned off the protons lose energy and return to the lower energy state; the time taken to achieve equilibrium is called the transverse relaxation, T_2. The rate of decay of the emitted signal is converted into a measure of the moveable fluids or free fluid index (FFI). Further processing can determine the volumes of irreducible and clay-bound water. NMR measurements can also be made on core samples in a laboratory to obtain calibration data to improve results.

3.4 Water saturation

Water saturation (S_w) is the measure of pore volume filled with water; the water may be mobile or capillary bound. Water saturation can be defined as effective or total depending on the porosity terminology being used. There are also a number of different terms to describe water saturation: initial (S_{wi}),

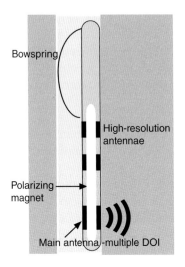

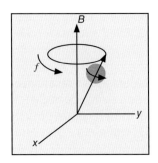

Above-In an external magnetic field, the precessional frequency (f) of a nucleus depends on the gyromagnetic direction (y) and the strength of the field (B)

A strong permanent magnet (B) induces alignment in the protons of hydrogen atoms in the reservoir fluid. Antennae monitor and record the electromagnetic signal of the protons as they precess back to their normal state of alignment with the Earth's magnetic field once the magnet is turned off: this is know as relaxation. The time taken for the protons to align is T1 and to relax is T2. Different companies use different methods to measure these events.

NMR logs give data on porosity and pore distribution (permeability) as well type and quantity of moveable fluids in the reservoir.

Vertical resolution/DOI: 15–60 cm/3–10 cm
Logging speed: 450–1000 m/h (tool dependent)

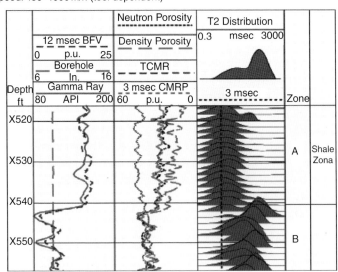

Figure 3.11   The nuclear magnetic resonance tool: principles of operation and typical display. (For a colour version see Plates section).

connate (S_{wc}) or irreducible (S_{wirr}). Initial water saturation is the proportion of water present in the pores at time of discovery; this may range from 100% to a value equivalent to the irreducible water saturation. Irreducible water saturation is the proportion of capillary-bound water that is immoveable by normal production processes. The irreducible water saturation varies with rock quality and will be higher in low-permeability rocks.

The determination of water saturation is a fundamental task in a study as it impacts on the overall distribution of fluids in the reservoir leading to an accurate calculation of the hydrocarbon volumes in-place. Unfortunately, this is more difficult to estimate than porosity because of the degree of uncertainty associated with the various measurements required. Water saturation, like porosity, can be an effective or a total property measurement, because it represents the proportion of the pore volume saturated with water and hence is a function of the porosity measurement and how that has been estimated and reported.

3.4.1 Core-derived water saturation

3.4.1.1 Fluid extraction
The routine core measurement of water saturation is carried out using a Dean–Stark apparatus in which an untreated sample is placed in a glass retort and heated at 105 °C for several hours to allow the trapped water to evaporate from the rock and be collected in a graduated chamber. Porosity is then measured on a similar cleaned sample and water saturation calculated as a function of the pore volume. The quality of the measurement can be extremely variable because the sample will have been affected by mud filtrate invasion during the coring process and by loss of fluids during the recovery of the core to the surface. Cores cut with oil-based mud should minimize the effects of invasion on the water saturation measurement.

3.4.1.2 Capillary pressure measurements
At its most basic, capillary pressure is the difference in pressure between coexisting pore-filling fluids. Together with wettability and relative permeability, it controls the rock–fluid interaction at the microscopic scale. The other forces acting on the reservoir fluids are gravitational and viscous, each of which combines to define the distribution of fluids in the reservoir.

The capillary pressure term (P_c) is used to define water saturation in a reservoir above a fixed datum known as the free water level (FWL) where P_c is zero and S_w is 100%. Figure 3.9 depicts a theoretical water saturation profile in a homogeneous sandstone reservoir. Because capillary pressure measurements are made in the laboratory, they require conversion to reservoir conditions; however, when properly handled they provide a wealth of data with which to characterize the most complex reservoir.

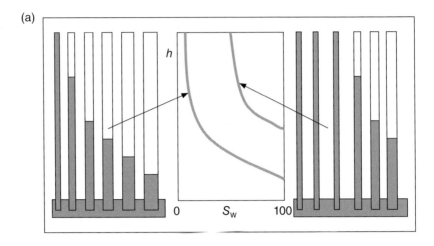

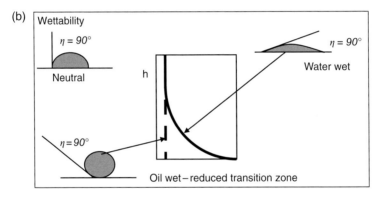

Figure 3.12 (a) Capillary pressure curves representing the effect of different pore-size distributions on fluid saturation. (b) The impact of wettability on saturation distribution with height above the free water level.

There is an inherent relationship between capillary pressure and water saturation in the reservoir because water is retained in the pore space by capillary forces. Capillary pressure can be expressed by the following equation:

$$P_c = \frac{2\sigma \cos\theta}{r}$$

where σ is the interfacial tension between the fluids and $\cos\theta$ is the contact angle between the fluids and the rock surface; r is the capillary radius or pore throat radius. In an oil reservoir, capillary pressure is zero and water saturation unity at the free water level; this is a physical datum that has more value than 'oil–water contact' when considering hydrocarbon distribution.

It is a simple exercise to convert laboratory measurements of capillary pressure to meaningful reservoir information using the density difference between the liquids and the height above the free water level:

$$P_c = Ch\Delta\rho$$

where C is a constant incorporating the gravitational term and dependent on the units being used ($C = 0.0069$ for field units), h is the height above the FWL and $\Delta\rho$ is the density difference between the wetting and non-wetting liquids.

There are a number of experimental methods used to determine capillary pressure in a sample (Tiab and Donaldson, 1996); the simplest and quickest is a destructive test using mercury injection (MICP). Mercury is injected into a clean, dry rock at predetermined pressures and at each step the volume of mercury injected is recorded. Because mercury is always a non-wetting phase, water saturation at different pressures can be calculated and related back to height above the FWL in the reservoir after a conversion for the different fluid systems. Because of the relationship to pore size distribution, MICP data are also used for rock typing exercises. Centrifuge and porous plate measurements of capillary pressure can also be made; these tend to be slower but more accurate. It is important to know what type of measurements have been made before comparing results to build a database

3.4.2 Wettability

Wettability is a measure of a rock's propensity to adsorb water or oil molecules on its surface in the presence of another immiscible fluid. It is important to understand the impact of wettability on the other dynamic properties of a rock as it influences fluid saturation and distribution in a reservoir. Rocks may be defined as water wet, oil wet, intermediate or mixed, depending on the results of relatively simple tests that measure the amounts of oil and water that are displaced from a native-state sample under certain conditions. While most (clastic) reservoirs would be considered to be water wet, under certain conditions all reservoirs can become oil wet at least in part. Carbonate reservoirs have a greater tendency for the oil-wet state because of the greater adsorption capacity of calcium/magnesium carbonate. High-asphaltene oils also have a greater tendency to adhere to the rock surfaces, thereby changing the wettability of a reservoir; that oil-wet carbonate reservoirs often contain high asphaltene oils is probably no coincidence. A final observation on the control of wettability is that high in an oil column, where the irreducible water saturation is least, there is an increased tendency for the oil-wet state. Many reservoirs are of mixed wettability: oil wet in the large open pores and water wet in the smaller isolated pores often filled with microporous clays.

3.4.3   Electrical measurements

To obtain a reliable estimation of water saturation requires a number of empirical relationships; these were established in a series of experiments undertaken by a Shell Oil researcher, Gus Archie, and published in 1942 (Archie, 1942). Archie used clean, clay-free, sandstone samples saturated with a brine of known resistivity (R_w) to estimate the *in situ* resistivity of the rock (R_o): he called this the formation resistivity factor (F):

$$F = \frac{R_o}{R_w}$$

He further demonstrated that there is a strong linear relationship between the logarithmic transform of the formation factor and the porosity of the sandstones:

$$F = \frac{1}{\phi^m}$$

where m, the slope of the line, had different values depending on the consolidation of the sandstone; the numerator was later generalized to the term a to accommodate different classes of sandstone (Winsauer et al., 1952). The terms a and m are known as the tortuosity factor and cementation factor, respectively.

In a fully saturated sample, the Archie equation becomes

$$S_w = \left(\frac{aR_w}{R_t \phi^m} \right)^{\frac{1}{n}}$$

The experiments made in the laboratory are designed to resolve the following elements of the Archie equation:

$$\text{Formation resistivity factor}: F = R_o \, / \, R_w = a \, / \, \phi^{-m}$$

and

$$\text{Resistivity index}: RI = R_t \, / \, R_o = S_w^{-n}$$

where R_o is the resistivity of a fully brine saturated sample, R_w is the resistivity of the saturating brine and R_t is the resistivity of the sample at different values of saturation.

In the first experiment, the sample is fully saturated with a brine of known salinity, electric current is passed through the sample and the resistivity is measured. By plotting F against porosity on a log–log plot for a number of similar rock types, it is possible to obtain the slope of the line, m, or the cementation factor (Figure 3.13). The value of m varies for different rock types as a function of the degree of cementation, ranging from <1.6 for poorly cemented rocks to >3.5 for very well cemented rocks; the default value for m is usually 1.8–2.2. The tortuosity factor, a, reflects the complexity of the connected pores and is usually set to 1.

The resistivity index is determined by measuring the resistivity of a sample at a number of different saturations, as the sample is slowly desaturated. At each saturation point, the sample is removed from the apparatus and weighed to determine the remaining saturation; this relates to R_t. The default range of values for n is also 1.8–2.2 (Figure 3.14). These tests may be performed under ambient or overburden correction using oil field or synthetic brine if a full characterization is required.

The Archie relationships were developed from clean brine-bearing sands; in a real reservoir, of course, the sands are likely to contain clays and hopefully oil to complicate the electrical system. Many different relationships have been developed for geographically specific areas or for special conditions; in all cases these should default to Archie when the sands are clay free and the formation water is significantly saline.

3.4.4 Log-derived water saturation

Resistivity logs are used primarily to distinguish water-bearing from hydrocarbon-bearing intervals, but can also indicate permeable horizons and estimate porosity. The only part of a formation able to conduct electricity is the water in the pore space or trapped by clays; the rock matrix and any hydrocarbons are normally resistive. Resistivity tools generate a current in the formation and measure the response of that formation to that current. The strength of the response varies with the salinity and volume of the formation water; more saline water gives a proportionally lower response than fresher water. Resistivity logs are usually presented in track 2 if in conjunction with a sonic log display or across tracks 2–3 of a standard API display; the scale is always logarithmic with a range of 0.2–200 or 2000 ohm metres (ohm m).

The resistivity of a porous rock depends entirely on the electrical conductivity of the formation fluid and mud filtrate, as the surrounding rock matrix acts as an electrical insulator. As previously described, drilling fluid can penetrate a permeable formation, forming a mudcake on the borehole wall and flushing the formation water away from the immediate surrounding volume, producing an annulus filled with mud filtrate; the depth of the annulus is a function of the permeability. The resistivity of the flushed or invaded zone depends on the resistivity and saturation of the mud filtrate (R_{mf} and S_{xo}) and any remaining formation water (R_t and S_w) and the porosity. When these

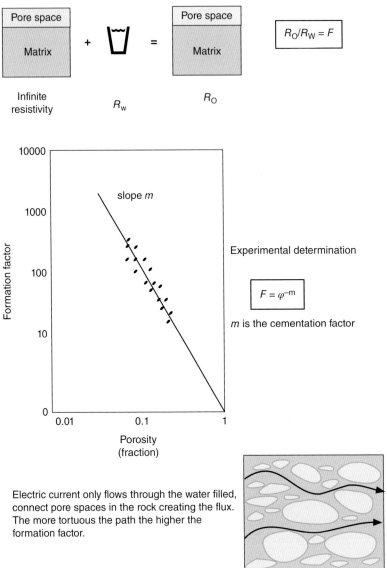

Figure 3.13 Formation resistivity factor (*F*): the principle of estimation and experimental determination. Plotting the results of each measurement determines a slope *m* that relates *F* to porosity, known as the cementation exponent. (For a colour version see Plates section)

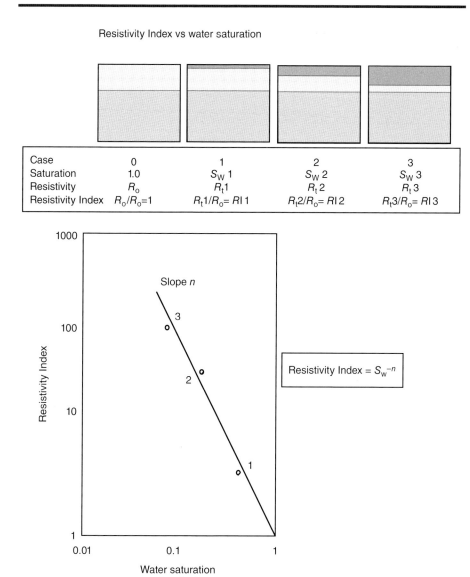

Figure 3.14 The resistivity index relates the proportion of a saturating fluid to the resistivity of a non-conductive fluid. (For a colour version see Plates section)

values are known, the resistivity in the flushed zone can be corrected for invasion. The resistivity of the uninvaded formation depends on the resistivity and saturation of the formation water and the porosity, where summing together the oil, gas and water saturation is unity (Figure 3.15). Typical values of R_t vary from 0.2 to 2000 ohm m and will only be investigated by the deepest, focused resistivity tools.

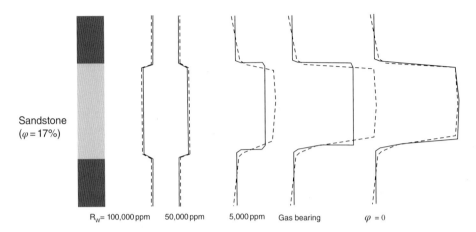

Sandstone
($\varphi = 17\%$)

$R_w = 100,000$ ppm   50,000 ppm   5,000 ppm   Gas bearing $\varphi = 0$

Figure 3.15 Resistivity log response in sandstone of constant porosity varying with formation water resistivity or hydrocarbon content.

There are two families of resistivity tools: electrode tools (laterologs) that have electrodes set on tool pads that produce a current and measure the formation response and induction tools that use electric coils to induce a current in the formation and measure the formation conductivity (Figure 3.16). The former can only be used in boreholes filled with a water-based mud; induction logs can work in both water- and oil-based mud systems. Resistivity is measured in units of ohm m and conductance as millimho/m. Resistivity tools were the first logging tools, developed in 1927 by the Schlumberger brothers, and there have been many evolutionary changes in their design, mode of operation and functions. Modern resistivity tools combine many electrodes and sensors to provide 360° borehole coverage at multiple depths of investigation. These 'array' tools are run in combination with most other formation evaluation tools to capture the key inputs needed for accurate estimation of hydrocarbon saturation.

3.5 Permeability

Permeability is defined as the ability of a reservoir to 'conduct' or 'transmit' fluids through the rock matrix: the flow capacity of a reservoir. While it is among the most important of reservoir properties to know, its measurement is also amongst the most difficult to acquire at the appropriate representative scale. Permeability is measured in darcies, reflecting the name of the person who first experimented with the flow of water through sand packs in 1856. Henry Darcy was a French municipal engineer based in Dijon for most of his career; he died in 1858, aged 55 years, just 2 years after completing his experiments in fluid flow (Figure 3.17). Darcy's simple empirical equation

DUAL LATERO LOG/ARRAY RESISTIVITY

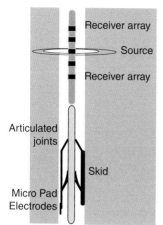

A low frequency, focussed current flows into the formation. Electrode arrays force the current in to a horizontal disk shape. The formation resistivity is determined by the amount of current flowing.

Vertical resloution/DOI: Deep-60cm/2–3 m
Medium-60 cm/60–100 cm

Logging speed: 1800 m/h; slower if run in combination with shallow reading log

Only run in open-holed filled with WBM. Tool must make electrical contact with the formation

Borehole and invasions required to estimate formation resistivity:
Deep~Shallow<R_{xo} indicates minor invasion
Deep<Shallow~Rxo indicates deep invasion

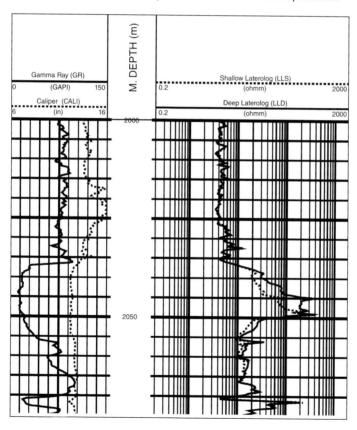

Figure 3.16 (*Continued*)

DUAL INDUCTION

Transmitter coils induce an alternating current in the formation. Receivers detect the magnitude and phase of the response, whichis proportional to the formation conductivity. Array tools have many receivers at small spacings, and use signal processing to create a single vertical resolution

Vertical Resolution/DOI: Deep-60 cm/1m: Medium-60 cm/120 cm: Shallow-<45cm/45 cm Logging speed: 600–1000 m/h depending on tool type;
Can be run in WBM or OBM; tools must be kept away from the borehole wall so have a stand-off assembly.

Must be corrected for borehole size,mud resistivity and bed thickness to estimate formation resistivity. Medium and shallow curves required for invasion corrections

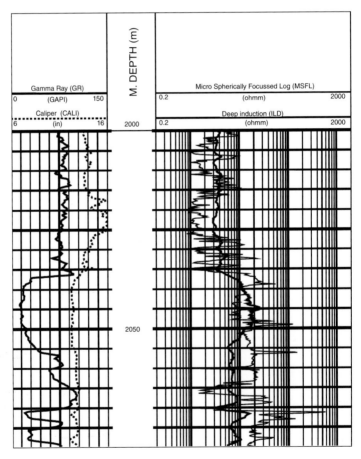

Figure 3.16   Different resistivity tools: laterologs and induction logs; modes of operation, application and typical displays.

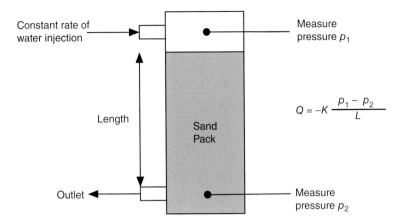

Figure 3.17 Experimental method to determine flow of water through a sand pack recreating the original Darcy experiments.

relates flow rate to a pressure differential across a sand filled pipe of known cross-section and length:

$$Q = -K \frac{p_1 - p_2}{L}$$

The constant K is the *absolute* permeability of the sand pack to a known fluid, water. Other restrictions are that the flow must be laminar or viscous and not turbulent, that there is no reaction between the fluid and the rock and that the fluid phase is unique and saturates the rock completely.

The more familiar equation used in core analysis experiments incorporates the cross-sectional area of the sample and also the viscosity of the liquid:

$$q = \frac{kA\Delta p}{\mu L}$$

where q = rate of flow, k = permeability (usually to air), A = cross-sectional area of the sample, Δp = pressure differential, μ = viscosity of the flowing medium (gas) and L = length of the sample.

3.5.1 Core permeability

Routine permeability measurements are made using the same core plug and flowing air or helium through the sample at a constant rate while varying the outlet pressure using different sized orifices. Measurements are made on both horizontal plugs (those drilled orthogonal to the core) and vertical plugs (drilled along the core); however, vertical measurements are usually only made on one-tenth of the number of overall samples. Permeability measurements can be routinely corrected for both gas slippage, the so-called Klinkenberg effect, and for overburden conditions. As with porosity measurements, there are measurement errors and sample volume/representation issues to contend with; the results for fractured samples should be ignored, for instance. The greatest issue with core permeability results is the matter of scale: the plug is taken to represent a homogeneous, one-foot interval of the reservoir that can be used to estimate permeability throughout the core and ultimately the well. Comparisons with log-derived permeability estimates and well test results regularly demonstrate the heterogeneous nature of reservoirs.

Relative permeability is the normalized value of effective permeability for a fluid to the absolute permeability of the rock. Relative permeability expresses the relative contribution of each liquid phase to the total flow capacity of the rock:

$$k_{ro} = \frac{k_o}{k}, k_{rw} = \frac{k_w}{k}, k_{rg} = \frac{k_g}{k}$$

where k_o, k_w and k_g are the effective permeability to each potential fluid phase. The measurement of relative permeability is fraught with difficulties and results must be treated with care. Wettability issues and small-scale heterogeneities in the sample affect measurements, and consideration must be given to these and other experimental issues when evaluating the results for use in dynamic simulation.

Relative permeability is measured in either a steady-state or an unsteady-state experiment. In a steady-state experiment, a fixed ratio of liquids is flowed through the sample until pressure and saturation equilibrium is reached; achieving steady-state flow can be time consuming, especially in less permeable material. The effective permeability of each liquid is calculated as a function of the relative saturation using Darcy's law, by measuring the flow rate, pressure differential and saturation. Monitoring the total effluent from a core sample during an imposed flood and calculating the relative permeability ratio that is consistent with that outcome is the basis of unsteady-state measurements. Steady-state experiments are more reliable and accurate, but take longer than the cheaper unsteady-state tests, which provide a greater interpretational challenge.

3.5.2   Log permeability

Permeability should only be calculated from logs when the formation is at irreducible water saturation. This condition can be determined using the bulk volume water (BVW) relationship:

$$BVW = S_w \times \phi$$

When the BVW values are constant, the interval is at irreducible saturation. A number of empirical relationships have been developed over the last 50 years based on laboratory experiments. The Coates–Dumanoir equation (Coates and Dumanoir, 1973) relates hydrocarbon density and porosity with resistivity measurements, but depends on two experimentally derived constants: these techniques are seldom employed today because of the uncertainty in the process. Whatever the approach taken to estimate permeability, the results must be compared with well test data to be considered meaningful.

Nuclear magnetic resonance provides the only means of estimating a continuous vertical permeability through a reservoir. The method depends on the relaxation time of the magnetized hydrogen atoms in the reservoir fluid and the relationship to pore throat size and hence permeability. There are two general types of empirical methods to use the data, one based on the irreducible water saturation (capillary-bound water) and the other using the transverse relaxation time (T_2); the commonest forms of these relationships were developed by Timur (1968) and Schlumberger–Doll Research (Kenyon et al., 1988), respectively. The use of either approach is best made after laboratory calibration on representative core samples.

Pressure draw-down and build-up measurements recorded during a wireline formation sample test both indicate the mobility of the fluids and formation being sampled, as does the rate at which the sample is recovered. These are only qualitative or at best semiquantitative estimates, but they are extremely valuable when comparing other results of permeability estimation.

3.5.3 Porosity–permeability relationship

It its simplest form, permeability can be predicted from the log–linear relationship with porosity determined from core analysis (Figure 3.18). Too often no more thought is given to the problem and only one relationship is propagated through the geological and petrophysical models. In reality, there is no causal relationship between porosity and permeability; rather, permeability is a function of grain size and sorting and the resultant pore throat size distribution. However, permeability can also be related to many other properties, either empirically or intrinsically, including pore surface area, irreducible water saturation, relative permeability and capillary pressure. In well log analysis, the only available predictor is porosity alone or possibly in combination with water saturation and volume of shale.

The first step in the workflow is to establish an empirical relationship between core-derived porosity and permeability, constrained by some zone (stratigraphic) or facies classification. The porosity data should be overburden corrected where possible and calibrated with liquid permeability results if possible. The data should be plotted with porosity on the *x*-axis and the logarithm of permeability on the *y*-axis; in this case the prediction will be a *y*-on-*x*

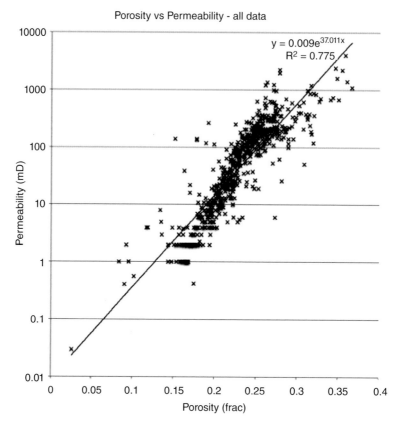

Figure 3.18   Example of a porosity–permeability cross-plot with a single linear *y*-on-*x* relationship described. The data distribution suggests that more than one lithofacies may be grouped together: try to partition the data to reflect geology.

relationship, that is, predicting permeability from porosity. The most useful form of the log–linear relationship is given by:

$$\log K = a\phi + b \quad \text{or} \quad K = 10^{a\phi + b}$$

where *a* and *b* are constants for each facies or reservoir zone. These simple relationships typically overpredict permeability at the low end and underpredict at the high end. It is usually better to develop a relationship based on some other property related to the lithology or rock type.

Relationships can be established on a field-wide basis or individual reservoir zone, facies or rock type over different porosity and permeability ranges. If a single empirical relationship holds for groups of facies then these can be combined if there is no inherent geological difference between them; sheet sands should not be combined with channel sands even if there properties are apparently the same because they will have different depositional character-istics that may control flow in the reservoir. It may be necessary to truncate the upper limit of predicted permeability so they do not exceed the maximum matrix or intergranular permeability measured on core for a particular facies or zone. This manipulation of results is not desirable but is often required to build the relationships required for subsequent property modelling.

3.5.4 Overburden correction and Klinkenberg effect

The old driller's adage that 'the only thing you know for sure about a core is that it used be in the ground' is especially true when one considers the changes that both rock and fluid undergo during coring, recovery and handling from rig to laboratory. The cutting of the core tends to introduce an invasion ring around the core that varies in thickness depending on the permeability. The release of pressure as the core is brought to the surface relaxes the grain fabric in most sandstones, increasing the pore volume and opening natural fractures – the opposite of rock compressibility. Water and oil will tend to seep out of the core, while gas may often be seen bubbling on the surface of a core on the rig floor immediately after recovery.

Porosity and permeability need to be corrected from laboratory (ambient) to reservoir (overburden) conditions; permeability is also often corrected for frictional slippage of the medium (gas) used in making the measurement, the so-called Klinkenberg correction.

The overburden correction is calculated after taking a series of measure-ments on a representative range of plugs at different confining pressures up to the calculated (*in situ*) net overburden pressure at the reservoir/sample depth. An empirical uniaxial correction is then applied to simulate the reservoir stress conditions. The change in porosity with applied confining stress is plot-ted on a series of normalized porosity reduction curves and the curve closest to the net overburden pressure is used to estimate the correction factor. Core compaction corrections are sensitive to the application of different simulated reservoir stress conditions.

The effective overburden stress is generally calculated as

$$\bar{S}_v = S_v - P$$

where S_v (total overburden stress) = depth × overburden gradient and P = initial reservoir pressure. Hence

$$\phi_u = \frac{\phi_a - U_c \times \Delta PV}{1 - U_c \times \Delta PV}$$

where ΔPV = change in pore volume, ϕ_a = ambient porosity and ϕ_u = uniaxial corrected porosity [assumed to be 0.62 after Teeuw (1971)].

To calculate the effect of uniaxial loading of the sample (rather than the confining pressure of the experiment), the Teeuw correction is applied. This correction is only valid in the case of linear elastic strain conditions and does not account for tectonic stresses. The correction factor is a function of Poisson's ratio of compressibility for solids.

Commonly, the correction factor will vary from 0.92 to 0.98 for most lithified sandstones; poorly consolidated sands will require lower values (or a completely different analysis method), while cemented sands will require less correction. Carbonates require little correction, as the rock fabric is often more robust.

The Klinkenberg correction is applied for gas flow under low pressures, as in the laboratory experiment, where the volume flux per unit area is greater than that seen at reservoir conditions. This effect is described by the following equation:

$$k_L = k_\infty \left(1 + \frac{b}{PM} \right)$$

where the constant b is the average capillary radius (pore-throat size) and the free path length of the gas molecules ($b = \lambda/r$). The PM term relates to the extreme value of permeability as it tends to infinity: this is a rock-specific parameter. Experiments have shown that the Klinkenberg effect is more pronounced when hydrogen is the flowing gas rather than nitrogen or carbon dioxide because of the size of the gas molecules.

3.6 Summary

In some ways, this chapter is the key to petrophysics as it deals with the fundamental properties of the porous medium and the contained fluids; but it is also one of the most confusing, because of the variety of ways to define and measure different properties. It is important to define clearly whether you are working in a 'total' or 'effective' porous system and how you might propose to move from one to the other. Most petrophysicists would insist on starting in the 'total' domain because this is the easiest to calibrate with core measurements, but of course our logs are working *in situ* in the reservoir, where our measurements will be essentially effective. The debate revolves around the volume of water associated with detrital clays or microporosity. To ignore this clay-bound water is to overestimate the hydrocarbon, a cardinal sin especially when calculating oil or gas initially in place in a reservoir model.

4
Quality Control of Raw Data

4.1 Validation of log data

The purpose of log data validation is to establish that the data loaded from the corporate database or digital tape archive represent what they are supposed to be and that they are numerically correct, precise and of usable quality for the purposes of the project. Ideally, the reference set of logs should be the paper copies of the field tapes; however, these original data are seldom seen these days and the default reference has become the corporate database.

Always question the data provided by any database; try to view all relevant data held in the database and select those which are most appropriate for your needs; it may not always be the most recent version. As most validation is done on digital versions of the data, it may be worthwhile displaying those logs that purport to be the same thing simultaneously and to look for discrepancies or deviations from the expected. Care should be taken to ensure that previously edited logs are not presented as the original data as provided by the service company; in this case, the potential for misidentification or data corruption is hugely increased.

Special care is needed where different logging runs cover the evaluation interval. If, for example, a proposed evaluation interval is covered by two dual induction log (DIL) runs, these should be treated as separate for validation purposes, even if they are subsequently merged into a single log curve for analysis. Always check which logging company acquired the data; operators will often switch companies after the exploration or appraisal phase of a development project and acquisition parameters may change, resulting in two or more sets of data.

Petrophysics: A Practical Guide, First Edition. Steve Cannon.
© 2016 John Wiley & Sons, Ltd. Published 2016 by John Wiley & Sons, Ltd.

If you are faced with a raw log database, which is often the case when starting a new project, the following workflow should prevent you falling into most of the traps that abound in data processing:

1. Load raw digital log data into the processing software and compare visually with field prints, especially if the logs have been digitized; an essential quality control step.
2. Collect all header information to identify tool type, borehole conditions and temperature and mud properties; determine top and bottom depth for each log run, especially if runs are to be merged.
3. When merging log runs, check for depth shifts; tie runs back to a casing point or marker level. Comparison of two log runs over the same interval is a simple quality control step.
4. Edit logs to remove erroneous data such as intervals of tool sticking or oversized/washed-out borehole: where gaps occur in reservoir intervals, try to calibrate reservoir properties from other logs.
5. Depth shift all logs to the chosen reference log, usually the GR from the resistivity log run or the density/neutron, as this is often the slowest logging suite run in most wells.
6. Depth shift core data to log reference usually with the core gamma log or the grain density data.
7. Perform environmental corrections; borehole size, temperature, pressure, mud properties and other factors that might influence the log responses

Most of the above can be done with modern software products; this is a time-consuming element of any project and can take up to a day per well to do it properly. A detailed discussion of aspects of data processing for interpretation follows.

4.1.1   Labelling

The purpose of this step is to check that the data in the various channels (columns) are what they claim to be and that they cover a sufficient depth range to include the proposed evaluation interval or reservoir unit. A common inconsistency occurs when the database includes data from different service companies; operator error is possible when a standard template is used to read the data without making provision for minor variations in data format between service companies. A typical error is made when the loader or user of the data mixes up different gamma ray traces. For future use in other software products, it is useful to be consistent in the naming convention for final logs.

The other major labelling issue is to verify that the log header information is complete; location, depth interval, elevation, casing depths, hole size, mud type, mud properties and temperatures are recorded along with logging tool data such as stand-offs and processing parameters (Figure 4.1) These data are required for environmental and tool corrections prior to log analysis.

```
~Version Information
VERS .          2.0              :CWLS Log Ascii Standard - Version 2.0
WRAP .          NO               :One line per depth step
~Well Information Block
#MNEM.UNIT      Data Type                 Information
#---------      ---------        -----------------------------------
STRT .M         295.00000        :START DEPTH
STOP .M         2519.00000       :STOP DEPTH
STEP .M         0.10000          :INCREMENT
NULL .          -999.25          :NULL DATA VALUE
COMP .          MYOIL            :COMPANY
WELL .          MYOIL-1          :WELL
FLD  .          MYFIELD          :FIELD
LOC  .          MYTOWN           :LOCATION
PROV .                           :PROVINCE
STAT .                           :STATE
CNTY .                           :COUNTY
CTRY .                           :COUNTRY
DATE .                           :DATE
UWI  .          0987654321       :UNIQUE WELL UWI
LIC  .                           :LICENSE NUMBER
API  .                           :API NUMBER
SRVC .          SCG              :SERVICE COMPANY
~Parameter Information Block
#MNEM.UNIT      Value                     Description
#---------      ---------        -------------
LONG .DEG       55.0463          :Well Surface Location
LATI .DEG       -3.9932          :Well Surface Location
XWELL .M        500456           :X-coordinate of Well Head
YWELL .M        5877421          :Y-coordinate of Well Head
~Curve Information Block
#MNEM.UNIT      API Code          Curve   Description
#--------       --------------    -----   ------------------------
DEPT  .M                          :0      Index
AC    .US/F                       :1      AcousticTravelTime
CAL   .IN                         :2      Caliper
GR    .GAPI                       :3      Natural Gamma Radioactivity
RT    .OHMM                       :4      Resistivity from Conductivity
SN    .OHMM                       :5      Short Normal
SP    .MV                         :6      Spontaneous Potential
```

Figure 4.1 Example of an LAS well file header showing the basic well information such as location and tools run. This example is of a composite suite of data for a complete well after well editing.

4.1.2   Parametric ranges

This stage of quality checking involves verifying that the data are plotted on the same scales and with the same numerical designators as the original data. For instance, it is possible that the neutron log has been recorded in sandstone matrix units instead of the default limestone units, even though the labelling indicates otherwise.

4.1.3   Repeatability

Repeatability is a basic measure of log quality; where a repeat log section has been run, it should be checked against the corresponding logs of the same interval in the primary log run. This obviously can be done digitally or using that rare commodity, paper data! Unless the original data are available, the user must trust that this fundamental quality control step was performed prior to loading into the database. This is an uncertainty in the data that must be accepted.

4.1.4   Tension

Review the tension log for each logging run and identify intervals of tool sticking; these will be clearly visible through pronounced tension departures from the gradual drift with depth. All such occurrences should be noted for use in further log quality control. For example, if the density–neutron log run is free from instances of tool sticking, then the associated gamma ray should be selected as the depth reference log, unless there is some other reason for not selecting it; perhaps that log suite investigated the evaluation interval over two runs rather than a single one. Modern logging operations often include single tool runs, so-called 'grand slams', which can work for or against the subsequent log quality control and interpretation: on the positive side, the single run will have a better chance of capturing all required data before hole condition deteriorates; however, if the hole is already unstable there is less chance to get comparative runs.

4.1.5   Borehole conditions

Calliper logs are run to measure borehole diameter, usually with density-neutron and resistivity tool runs; choose the former where possible. Borehole rugosity can be measured with the density correction log (Drho) associated with density logs as an additional quality control tool. These logs are used to identify intervals where the borehole size increases more than 2 inches beyond the nominal drill-bit diameter; beyond this increase, the density log will indicate progressively more optimistic measurements of porosity. Additionally, a Drho correction of greater than $\pm 0.15\,g/cm^3$ also indicates intervals of unreliable density log measurement. All such intervals should be flagged and replaced with alternative porosity interpretations.

4.1.6 Noise spikes and cycle skipping

Noise spikes and cycle skipping are observed only on sonic tools. Tool centralizers rubbing against the borehole wall generate noise spikes. A noise pulse can be triggered in a receiver circuit ahead of the real arrival pulse, creating a spike of low transit time. Tool centralizers are not always employed, so a check of the log header will provide the necessary information.

Cycle skips occur when the first negative arrival of the signal, usually at a far receiver, falls below the trigger level. This can occur at one or both of the far receivers; the effect is to make the log reading 10–20 ms/ft too high, depending on the receiver. These artefacts are usually localized but can be extensive under certain circumstances, such as unconsolidated gas sands where the level of signal attenuation is high. Identify artefacts of both types for later attention; they may be removed through editing, but this should be done after depth shifting is completed, as these data may be useful in correlating borehole conditions.

4.1.7 Editing log data

There are three likely reasons for editing a single log or a suite: sticking tools, poor borehole conditions and noisy data. Logging tools that require contact with the borehole wall are the most likely to stick; the density–neutron combination has a sprung calliper arm on the density pad, for instance. This may have an influence on which gamma ray should be used as a depth reference. Edited logs should be clearly identified in the database and a record made of the history of the edits.

4.1.8 Creation of pseudologs

Sometimes, it is necessary to create a log over a section of a well using another log as source data. This may happen in the case of a pseudo-sonic log, required for geophysical modelling, based on an inverted density log. If the sonic log is very noisy over a particular borehole section, the density log, if reliable, can be used to develop a simple relationship to predict the sonic response. In the case where the gamma ray from one log run is more reliable than another one because it was logged at a slower speed, then rather than just splicing in the preferred log, a calibration between the two curves should be made.

4.2   Depth merging

Depth merging of logs is one of the most time-consuming elements of the quality control and data preparation process. It cannot be automated successfully and requires patience and accuracy by the interpreter to achieve consistent results; if done properly (or improperly), this process can influence all the subsequent aspects of the well evaluation. There are three steps in the process; in all cases it will be necessary to perform the first two steps:

1. Depth shifting of a benchmark log, usually the gamma ray.
2. Applying these shifts to all logs recorded with the corresponding gamma ray.
3. Further shifting of logs as required.

All depth shifts should be checked against the final paper prints of the logs or the digital data tape from the wellsite; these should be taken as the definitive version, not an edited file from a database. Occasionally, it will be apparent that logs recorded by a single logging tool combination will not be reconciled with each other in terms of depth. Minor differences can be checked against the paper logs and, if still apparent, they should be left alone as they probably represent the consequences of different tool response functions. In the case of a catastrophic failure of the tool memory, for instance, there may be a discrepancy of up to 12 ft between any two digital readings with the same tool string, even though paper logs indicate correct depth matching.

In all situations, it is necessary to select a depth reference log; usually this is the gamma ray run with the density–neutron tool combination, as this is commonly run at a slower speed than other tool combinations. If the density–neutron combination tool string indicates tool sticking during the run, then an alternative depth reference log may be required. One of the values of returning to the original data is to appreciate comments made by the logging engineer regarding the conditions under which a particular tool string was run. It is always sensible to create, identify and store edited logs separately from the original raw data; a simple suffix to the log name will be sufficient to indicate that the log has been shifted.

4.3   Tool corrections

Wireline logs need to be corrected for environmental conditions and also for mud invasion. Most software products have the full range of tool correction charts available, but new tools require new correction charts, and these take time before they can be included as a software update.

4.3.1 Environmental corrections

Tool environment corrections compensate for the differences between the actual borehole conditions and the test pit where the tool was calibrated. The corrections are computerized versions of the departure curves and correction charts used by service companies and are therefore tool specific; there are no generic corrections. In general, the corrections compensate for borehole size, mud weight, temperature and other environmental conditions that might affect the logging tool measurement. These corrections cannot compensate for geological variations such as bed thickness or shoulder effects. Further corrections to resistivity measurements due to invasion effects can be made after the basic borehole corrections.

Environmental corrections should be performed after depth shifting and log editing. The exact depth range and tool combinations should be recorded; where possible, each logging run should be treated as a different log for the purpose of environmental correction. If the depth ranges for different runs of a given log are mutually exclusive, the corrected data log can be spliced together to form a composite log without affecting the original data.

Corrections should be applied to the gamma ray, density, neutron, dual induction, dual laterolog, microspherically focused log or proximity log and the spherically focused log. The different outputs of the spectral gamma ray tool, SGR (total GR) and/or CGR (i.e. total GR with uranium removed), may be corrected in the same way as the normal tool. The lithodensity (LDT) log correction will also correct the PEF curve. In the case of the dual resistivity tools, both deep and shallow searching logs require correction.

Ultimately, the depth-shifted, edited and borehole-corrected logs are the data that will be used for our interpretation; these logs should be identified as such in the project database.

4.3.2 Invasion corrections

Resistivity logs need a further correction to compensate for the invasion of water-based drilling mud filtrate into permeable formations; no corrections are required for induction logs that are run in oil-based mud. Invasion corrections are made after borehole environmental corrections. The correction procedure requires three input logs in one of the following combinations:

1. LLD, LLS and a micro-resistivity log, usually the MSFL.
2. ILD. ILM and a micro-resistivity log, usually the MSFL.
3. ILD, ILM and SFL.
4. ILD, SFL and a micro-resistivity log, usually the MSFL.

The commonest combinations are **1** and **2**. After the appropriate corrections have been applied, the following curves are generated:

- true formation resistivity – R_t
- flushed zone resistivity – R_{xo}
- diameter of invasion – DI.

4.4   Core analysis data

The purpose of core data validation is to ensure that the data about to be loaded into a project database are correct; correctly labelled, numerically correct, precise and of good quality. There is little point in loading incorrect data into a database as it will corrupt the results of any analysis or bring those results into question. The definitive record is that of the core analysis laboratory; however, even these data should be inspected for anomalies, artefacts and observations made at the time of acquisition (Figure 4.2). Core analysis data can come in a variety of forms, including continuous data from core gamma ray measurements and probe permeametry, to discrete values from core plugs and whole-core analysis. The discrete data are the most common types of core data used in reservoir characterization, whereas the continuous data are used for core-to-log depth correction or calibration of other measurements.

4.5   Merging core and log data

If core data are to be used for log calibration, it is essential that the two sets of data be exactly on depth. There are usually a number of methods by which this can be achieved, either using a core gamma log or actual core analysis results, usually measured porosity or grain or bulk density values. Both methods are highly subjective and dependent on the completeness of the core and any core handling errors that may have been made at the wellsite or subsequently.

Matching core gamma and log gamma ray data is usually performed with digital records of both curves displayed in a common window in the same way that other logs are depth shifted. The log data should be the definitive curve against which the core is shifted up or down as required. Shifting should always be done on a core-by-core basis, especially where there is any evidence for missing sections of core, extensive rubble zones or indication of poor core handling. When there is no digital record and the matching is done on paper records only, a series of paired points are recorded for each section of core.

SCG LABORATORIES UK LIMITED
RESERVOIR EVALUATION STUDIES

PAGE 1 OF 6

COMPANY: MY OIL LTD
WELL: MY OIL-1
FIELD: MY FIELD
LOCATION: MAELOR

DATE: Oct-15
FORMATION: BIRTHDAY
DRLG FLUID: OBM
ELEVATION: 103M

FILE NO:03101955
ANALYSTS:SJC

SAMPLE NUMBER	DEPTH	HORIZONTAL PERMEABILITY	VERTICAL PERMEABILITY	HELIUM POROSITY	OIL SATN	WATER SATN	GRAIN DENS	DESCRIPTION	COMMENTS
CORE 1									
1	3522.25	0.04		3.7	10.3	76.9	2.68	SST, WH-LT GY, VFGR, W STD, CALC CMT	LOW GRAIN DENSITY MAY INDICATE INCOMPLETE CLEANING
2	3522.50			11.0	19.2	64.9	2.61	SST, WH-LT GY, FGR, W STD, CALC, MICA	
3	3522.75	0.08	0.07	13.0	0.0	83.4	2.71	SLST, LT GY, CALC, ARG, CALCT XLS	
NPP	3523.00							SCAL	NO PLUG POSSIBLE
NS	3523.25								PRESERVED SAMPLE
4	3523.50	0.11		6.3	11.8	67.5	2.66	SST, LT GY, F-CRS, PSTD, CALC CMT, HD	
5	3523.75	0.26		5.4	15.3	36.3	2.68	SST, WH-LT GY, FGR, SL PBL, CALC, WCMT	
6	3524.00	0.04		2.0	0.1	55.5	2.67	SST, LT GY, F-CRS, PSTD, CALC CMT, HD	
7	3524.25	0.08		3.1	0.1	39.5	2.67	AS ABOVE	
NS	3524.50							SHALE	NO SAMPLE TAKEN
NS	3524.75							SHALE	NO SAMPLE TAKEN
8	3525.00	1.3		14.4	0.1	84.4	2.66	SST, LT GY, FGR, WSTD, CALC CMT, ARG	
9	3525.25	0.28		12.5	0.0	84.3	2.69	AS ABOVE	
10	3525.50	0.14		12.8	0.0	88.4	2.68	AS ABOVE	
11	3525.75	0.57	0.34	17.2	3.8	84.7	2.69	SST, WH-LT GY, FGR, W STD, CALC, MICA	
12	3526.00	0.07		11.6	0.0	93.9	2.66	AS ABOVE	
13	3526.25	0.8		12.7	0.0	95.1	2.70	SST, LT GY, VFGR, WSTD, CALC CMT	
CORE 2									
14	3530.50	6.6		21.1	0.1	86.8	2.67	SST, GY, VFGR, WSTD, CALC CMT, ARG, MHD	
15	3530.75	15		10.2	0.0	89.9	2.70	SST, GY, VFGR, WSTD, CALC CMT, SL ARG	
NS	3531.00							SCAL	
16	3531.25	14.8		15.6	0.0	83.7	2.70	SST, GY, F-CGR, PSTD, CALC CMT, FRAC	
17	3531.50	9.5		22.2	4.2	73.3	2.66	SST, GY, VFGR, WSTD, CALC CMT, SL ARG	
18	3531.75	20	1.7	23.1	11.9	81.2	2.65	AS ABOVE	

Figure 4.2 Example of a conventional core analysis data sheet. The Comments column may give clues to the quality of a particular measurement.

4.6   Converting measured depth to true vertical depth

After all the corrections, edits and shifts have been made, both log and core data are now referenced to an absolute value that is a measured depth starting at a datum based on the drill floor or rotary table elevation and increasing positively along the borehole. At some stage, this depth will need to be converted to a true vertical depth referenced usually to sea level or possibly ground level; the log headers or mudlog will provide the definitive TVD datum. This is usually done automatically based on a deviation survey comprising measured depth and inclination and azimuth of the borehole; deviation surveys can be acquired at a number of levels in the well, and the more data points are available for correction the more accurate the conversion will be. Sometimes, it is necessary to generate a simple deviation survey, perhaps referenced to geological layers. There are two main methods used to convert measured depth (MD) to true vertical depth (TVD) – cubic spline and minimum curvature; the results will be different, so it is important to know which has been used in the conversion. In the cubic spline method, depth is referenced to TVD, whereas the minimum curvature method uses the MD as the reference, so if the deviation survey is based on one method, ensure that that when converting a log the same method is used.

4.7   Summary

Garbage in – garbage out!

5
Characteristic Log Responses

A fundamental task in log analysis is to characterize the different log responses to the constituent rock matrix, clays and pore fluids found in the reservoir. There are numerous ways to approach the task; however, they all revolve around the distribution of the properties in stratigraphic space. Although it would be nice to think that a single set of shale points or matrix density would characterize the complete reservoir, this is unlikely; the partitioning of these characteristic responses is a function of the geology of the reservoir. The simplest way to do this is to zone the reservoir into geologically sensible layers and to determine the characteristic responses per zone. With respect to the gamma ray response to shale or sand, this is unique for every well; one way to consolidate the data is through normalization of the response. However, always be aware of spurious peaks or higher than expected responses due to organic-rich components, potassium feldspar or heavy minerals.

5.1 Characteristic shale response

The first step in the petrophysical interpretation workflow is usually to establish the volume of shale in the reservoir. The term shale has many meanings: to a geologist, it is a lithified, fissile mud rock produced by low-grade metamorphism; to a petrophysicist, it is the part of a reservoir that does not contribute the volume of hydrocarbon and can be detrimental to productivity. 'Shale' in the reservoir has two main origins, either detrital or authigenic, which is really clay. Detrital shale comprises mud and silt-grade quartz rocks deposited in association with the sands forming the reservoir and will impact on the porosity; authigenic clays are created by the post-depositional diagenetic processes acting on the reservoir and generally have a negative effect on permeability (Figure 5.1).

Petrophysics: A Practical Guide, First Edition. Steve Cannon.
© 2016 John Wiley & Sons, Ltd. Published 2016 by John Wiley & Sons, Ltd.

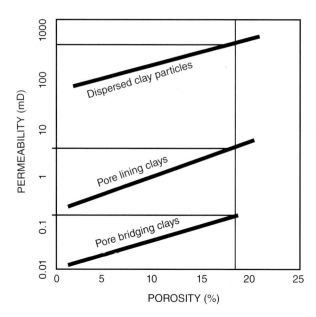

Figure 5.1 Schematic representation of the impact of different clay types on permeability in a clastic reservoir.

Detrital shale is found as structural, laminated or dispersed forms, related to the mode of deposition of the reservoir. Structural shale occurs commonly as shale clasts, often reworked into channelized deposits in fluvial or deep marine settings. As such, they occupy a volume of reservoir that otherwise could have contained sand grains with interparticle pores. Laminated shales are found interbedded with sands and will again reduce the gross reservoir volume; they are commonly associated with low rates of submarine deposition. Dispersed shales are often recognized in shallow marine, bioturbated sandstones where initial thin clay and silt laminae are reworked by the fauna living in the sediment.

Authigenic clays form within interparticle pore space, usually as a product of the dissolution of less stable mineral grains and lithic fragments. These clays form as pore-lining crystals, pore-blocking booklets and pore-bridging filaments; pore-blocking clays will reduce porosity and pore-bridging filaments will reduce permeability. Clays are hydrous aluminium silicates that have a large surface area, attracting and holding water molecules to the crystals by weak electrochemical forces; the water molecules are in continual motion binding and unbinding constantly. The presence of clays in the reservoir can have a deleterious effect on hydrocarbon production; special note should be made of the effect of different drilling and completion fluids. Common clay minerals are listed in Table 5.1; although chlorite is not strictly a clay, it is often included as one.

Shales and clays affect wireline logs in a number of ways and the tool responses and interpretation methods must be corrected for their effect. Gamma ray logs

Table 5.1 Common clay types and their characteristics.

Clay type	Habit	Composition	Reservoir problems
Kaolinite	Booklets	$Al_4(Si_4O_{10})OH_8$	Fines migration
Smectite	Pore lining	$(Ca,Na)(Al,Mg,Fe)_4(Si,Al)_8O_{20})OH_4$	Water sensitivity, microporosity
Illite	Pore bridging	$KAl_4(Si_7AlO_{20})OH_4$	Microporosity, fines migration
Chlorite	Pore lining, bridging	$(Mg,Fe,Al)_{12}(Si,Al)_8O_{20}(OH)_{16}$	Acid sensitivity
Mixed layer clays	Pore lining, bridging	Illite/smectite, chlorite/smectite, Chlorite/vermiculite	Water and acid sensitivity

are affected by the presence of radioactive elements, usually uranium associated with organic matter, and by high concentrations of heavy minerals and potassium-rich clays. Resistivity logs record a lower response in shaly sands, usually those with abundant authigenic clay, because of the greater conductivity of the clay-bound water associated with the clay and also the presence of microporosity; this results in an overestimation of the water saturation. The presence of shales and clays generally causes the sonic, density and neutron tools to record too high a porosity; the matrix velocity of shale is usually less than that of the reservoir sands, resulting in a higher calculated porosity; the neutron tool responds to the hydrogen atoms in the bound water molecules, again resulting in a higher porosity value. The density tool calculation also requires a correction for the generally lower shale density; however, where the shale and matrix density are the same, the density log will tend to give an accurate effective porosity.

Having established what type of shale and clays we are dealing with, it is possible to develop a method to discount their presence and effect on the subsequent interpretation. When dealing with detrital shale, the assumption is made that any shale bounding the reservoir zone is the same as the shale distributed with the reservoir, and this may not be the case, of course; however, the characteristic gamma ray response is used to determine a proportional volume of shale in the reservoir. A simple histogram (Figure 5.3) display of the interpretation interval including the bounding shale will give a shale maximum and minimum or 'clean' value for clean sand; when selecting the end members, exclude and outliers or 'tails' to the data. The maximum and minimum values can then be entered in a simple linear equation of the following form:

$$V_{sh} = \frac{GR_{log} - GR_{sand}}{GR_{sh} - GR_{sand}}$$

This relationship is also known as the gamma ray index (GRI) and is effectively a normalization of the tool response. This is an empirical relationship and

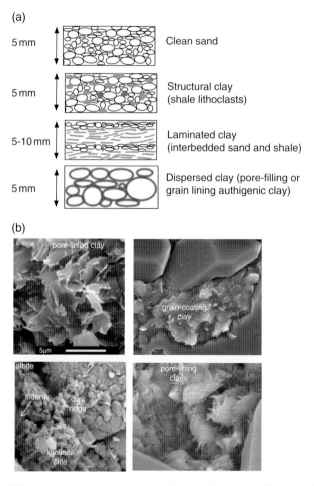

Figure 5.2   (a) Different clay types based on the Thomas–Stieber classification; (b) examples of different clays under SEM; pore-lining and grain-coating examples are seen.

there is reason to believe that the linear relationship is not true everywhere: different workers have developed many variations for specific geographic or stratigraphic settings.

The linear equation and its variants work well for structural and dispersed shales or where the shale type is not well determined. An alternative that can be used for dispersed and laminated shales is the Thomas–Stieber method that uses a relationship between porosity and normalized gamma ray for determining volume of shale:

$$GR_{lam} = \left(GR_{sand} - V_{sh}GR_{sand}\right) + V_{sh}GR_{sh}$$

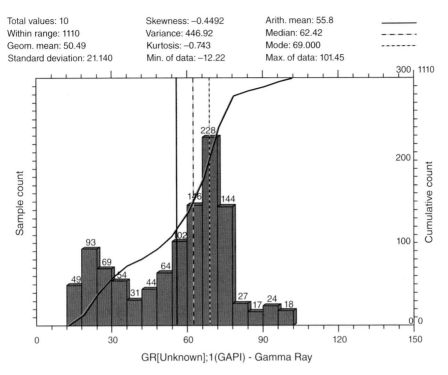

Total values: 10
Within range: 1110
Geom. mean: 50.49
Standard deviation: 21.140

Skewness: −0.4492
Variance: 446.92
Kurtosis: −0.743
Min. of data: −12.22

Arith. mean: 55.8
Median: 62.42
Mode: 69.000
Max. of data: 101.45

Figure 5.3 Histogram of gamma ray response in a well showing a bimodal distribution reflecting the two populations of sandstones and shales. This display can be used to identify the clean sand and shale responses for calculation of shale volume.

and the corresponding porosity relationship is

$$\phi_{lam} = \left(\phi_{sand} - V_{sh}\phi_{sand}\right) + V_{sh}\phi_{sand}$$

The Thomas-Stieber method has the effect of removing the influence of detrital shale on the subsequent porosity and saturation interpretations. In effect, this also reduces the gross reservoir interval also, but should increase the average porosity of the interval.

Because the gamma ray tool response is influenced by numerous borehole conditions, it is necessary to carry out the estimation of shale volume on a well-by-well basis; the maximum and minimum values will be well specific. If the spectral gamma ray log is available field-wide, the computed or corrected gamma ray curve (CGR) should be used for shale volume estimation because this curve has had the uranium component removed, leaving only the potassium- and thorium-related response.

Having established the 100% shale response, the next step is to identify the other log responses that correspond to it. One way to do this is by a series of cross-plots of V_{sh} against density, neutron and sonic logs for each interpretation interval. The bulk trend of the data is projected from lower values of V_{sh} to $V_{sh} = 1$ and an estimate is made of the intercept in each case. Where there is any ambiguity in the data, it is conventional to select the top of the data cloud for positive gradients and the base of the cloud for negative gradients. The resulting values are known as the shale points for the different logs. In complex lithologies and heterogeneous or stratigraphically complex reservoirs, it may not be possible to identify field-wide parametric shale values, in which case it will be necessary to determine well- or zone-specific values.

This process is applicable to most other logs; however, there are two further approaches that may be applied to specific shale characteristics: shale resistivity and a combined neutron–density approach. In fresh water systems, whether they comprise mud filtrate for S_{xo} determination or formation water for S_w calculation, the value of R_{sh} is crucial. By cross-plotting the appropriate $V_{sh} > 0.9$, if there are sufficient points, against an environmentally corrected deep resistivity log, any variation in R_{sh} in the well can be determined. If this variation is significant, it will be necessary to plot the logs of V_{sh} and deep resistivity along the borehole to identify separate zones where R_{sh} varies with depth.

It is possible to use the neutron–density combination as an alternative shale indicator (Figure 5.4), resulting in greater internal consistency of shale point identification. This approach can also be applied to the PEF, but should not be used for any of the other logs. First calculate a density porosity using default matrix (2.71 g/cm³) and fluid (1.0 g/cm³) properties for the water leg; if fluid density is known to vary through the well because of hydrocarbons, it may be necessary to build up the porosity data on a zone basis. The computed density porosity is an 'apparent' porosity log as it is not shale corrected. The next step is to plot the corrected neutron log (CNL) against the apparent porosity log. The neutron shale point should be obvious from the plot; however, the density porosity shale point is determined from the equation

$$\text{Den}_{sh} = \frac{\rho_{ma} - \rho_{sh}}{\rho_{ma} - \rho_{fl}}$$

where ρ_{ma} is the matrix density of limestone, ρ_{sh} is the density log value from the cross plot and ρ_{fl} is the appropriate fluid density.

Finally:

$$V_{sh_{N-D}} = \frac{a - b}{c - d}$$

where

$$a = \left(\rho_{clean} - \rho_{fl}\right) \times \left(CNL_{fl} - CNL\right)$$
$$b = \left(\rho_b - \rho_{fl}\right) \times \left(CNL_{fl} - CNL_{clean}\right)$$

$$c = \left(\rho_{\text{clean}} - \rho_{\text{fl}}\right) \times \left(CNL_{\text{fl}} - CNL_{\text{shale}}\right)$$

$$d = \left(\rho_{\text{shale}} - \rho_{\text{fl}}\right) \times \left(CNL_{\text{fl}} - CNL_{\text{clean}}\right)$$

5.2 Matrix characteristics

The evaluation of matrix characteristics should be carried out on a multi-well basis for each reservoir zone that is to be interpreted. Identification of the zones can be guided by grain density results from core analysis: if the variation in grain density, displayed as a histogram, is small for each zone, then the reservoir interval is likely to be homogeneous and the mean grain density can be used as characteristic for that zone. If matrix values are very variable, it may be necessary to adopt a more flexible approach, perhaps working on a well-specific basis or moving to a rock-type approach (see Chapter 9).

Matrix values should only be identified on clean intervals, $V_{\text{sh}} < 0.1$. A cross-plot of density–neutron logs over each interpretation interval will describe clusters of points about the matrix endpoints for different lithologies (Figure 5.4); if the matrix values have been determined on a well-by-well

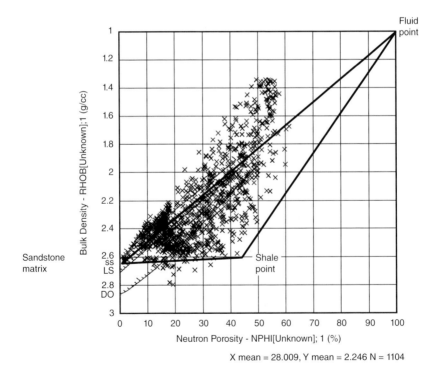

X mean = 28.009, Y mean = 2.246 N = 1104

Figure 5.4 A neutron–density cross-plot scaled to identify the sandstone, shale and fluid points of interbedded, water-filled sequences.

basis, then this plot should also be well specific. Where there is consistency, use a field-wide value for each interpretation zone where possible to improve workflow efficiency. The same procedure should be adopted for the neutron–sonic combination also, using the previously determined neutron matrix value to fix the sonic matrix. Where the PEF is available and is to be used in subsequent lithology evaluations, the matrix properties can be taken from tables.

5.3   Fluid characteristics

Because of mud invasion, the fluids present in the pores are seldom the original contents, but an unknown mixture of mud filtrate and formation fluid. As a result, the response of a shallow sensing tool to this fluid mixture must be determined empirically; this can only be done accurately over cored intervals where the matrix properties are known unambiguously. The method is usually applied only to the sonic and density tools but may be extended to the neutron and PEF logs. However, the procedure to determine these fluid characteristics is ambiguous and the results may be grossly misleading.

Starting with the sonic log, a fluid response is required to compensate for the deficiencies of the time-average equation when it is used in combination with porosity. The method is a cross-plot of sonic against porosity for intervals of consistent matrix values. A straight is line drawn through the $\phi_{sonic} = 0$ and the sonic matrix (Δt_{ma}) value; the resulting fit defines a value for sonic response in the unknown fluid mixture (Δt_{fl}). In general this will be somewhere between 185 and 240 µs/ft; if all else fails, use the default of 189 µs/ft. The same procedure can be followed for the density log, setting the fixed point at $\phi_{den} = 0$ and matrix density (ρ_{ma}).

5.4   Hydrocarbon corrections

The presence of hydrocarbons has an effect on all three main porosity tools; most interpretation software tools have in-built correction routines to compensate and these should be used.

In the presence of hydrocarbons, the interval transit time of a formation is increased, resulting in too high a calculated porosity. Hilchie (1978) proposed the following corrections: for gas $\phi = \phi_{son} \times 0.7$ and for oil $\phi = \phi_{son} \times 0.9$.

The density log response is affected in the same way as the sonic log, resulting in an overestimated porosity; the effect of oil in the pores is minimal, but the presence of gas can be significant. The response of the neutron log where the pores are filled with gas results in lower porosity estimations than the actual formation porosity. This occurs because in a fixed pore volume there is a lower concentration of hydrogen in gas than oil or water; this is not accounted for in the tool processing. This is the so-called 'gas effect' seen in wells with a gas cap. It can be corrected but usually is ignored and an alternative porosity measurement is used in gas-bearing intervals.

5.5 Shale corrections

The basic porosity tool responses can now be corrected for the effects of shale; this is done using the generic equation

corrected log = input log + V_{sh} (matrix response – shale response)

The input log should be the hydrocarbon-corrected version of whichever log-ging tool is being used. For example, the correction to the sonic log would be

$$\Delta t_{corr} = \Delta t + V_{sh}\left(\Delta t_{ma} - \Delta t_{sh}\right)$$

where Δt_{corr} is the shale-corrected sonic log, Δt is the depth-matched sonic log (no environmental corrections are applied), Δt_{ma} is the matrix transit time and Δt_{sh} is the shale transit time. For the neutron and density logs, the hydrocarbon-corrected inputs should be used. Once these corrections have been made, even though they may be imperfect, the resulting logs should be considered as tools run in a shale-free environment and no further reference need be made to shale effects other than in the selections of net pay cut-offs. Resistivity logs are not corrected for shale effects in this way; rather, corrections are made when performing the water saturation interpretation.

5.6 Summary

The key to this part of the workflow is to recognize that different intervals may have different rock or fluid properties and to describe adequately the differences as expressed by the log responses. The distinction is subtler than for sand or shale limits defined by the gamma ray maximum and minimum values: rather, it is the response of the deep, medium and shallow resistivity in sands of varying permeability. These distinctions will be followed through in the Archie parameters, helping to sample and test the full range of possible rock types. Changes in mineral content affecting the log responses can be used to determine different depositional environments; glauconitic sands are typical of shallow marine environments, coals are associated with floodplains and lagoon swamps and feldspathic sands are more commonly found in aeolian environments.

6
Evaluation of Lithology, Porosity and Water Saturation

Before estimating porosity or water saturation, reservoir lithology and formation water resistivity must be known. Ideally both will be known from core or cuttings and fluid samples; however, neither will be fully characterized in the early stages of a development. Log data can be used to distinguish gross lithology as part of a porosity interpretation and there are a number of ways to estimate R_w using the range of resistivity tools available.

6.1 Evaluation of lithology

The requirements of a lithology interpretation differ for the geologist and pet-rophysicist: the geologist is looking for information on depositional environment whereas the petrophysicist requires input for an equation; one is descriptive and the other is numerical. Both disciplines use core and cuttings data to calibrate their log interpretations, but the geologist will be looking for metre- to decimetre-scale patterns, whereas the petrophysicist requires values at every measurement interval of the logs. On the subject of scales of interpretation, remember that the vertical resolution of wireline logs is much coarser than for core and that the petrophysicist's task is to identify any layer that might contribute to hydrocarbon production. A different approach to data acquisition and interpretation is required when dealing with thinly bedded reservoirs.

We already know that there are two major lithological reservoir classes: clastics and carbonates; all other reservoirs, such as basalts and igneous basement, require special characterization. When scanning a suite of logs, it is worth remembering that the gross character of a sequence of clastics or

Petrophysics: A Practical Guide, First Edition. Steve Cannon.
© 2016 John Wiley & Sons, Ltd. Published 2016 by John Wiley & Sons, Ltd.

carbonates is generally very different (Figure 6.1). Shales of near-constant resistivity usually surround sandstone reservoirs, such that changes in the reservoir fluid are easily observed; a carbonate reservoir, however, will show rapid resistivity changes due to the variable rock types. In this section, we focus primarily on clastic reservoirs; Chapter 9 reviews carbonate reservoir characterization.

In the previous chapter, we saw how a shale volume log can be calculated from the gamma ray; but even before this, modern log analysis software allows the user to display the data in meaningful ways to help with the basic lithology

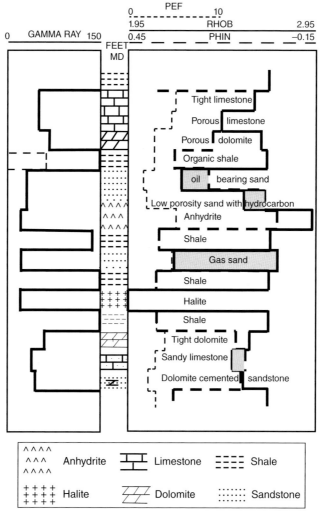

Figure 6.1 Simplified lithology determination from gamma, neutron–density and PEF logs.

interpretation. A simple colour-fill display of gamma variation will pick out the gross lithological variation and help with the identification of larger scale depositional patterns. Combined with a neutron–density colour fill where the two logs 'cross over' will identify the higher porosity layers in the sequence. These tools become important when selecting bed boundaries for subsequent analysis. It is convention that such boundaries are picked on the 'point of inflection' of the chosen log, i.e. the mid-point between the peak and the trough (Figure 6.2).

The next level of graphical display, made especially powerful since the development of computer software, is the semiquantitative interpretation of log data in the form of histograms and scatter plots to describe both the statistical distribution of properties and their inter-relationships. A number of different interpretive plots have been developed over the years to identify discrete mineral combinations that make up the constituent parts of a lithology: M–N plot (Figure 6.3), MID plot, LDT plot. Taken to the extreme, these methods result in a purely statistical lithology interpretation with all the hazards that might involve; however, these displays are valuable in understanding the relationships between lithologies and logs, hopefully with sufficient hard data from cores and cuttings to calibrate the results.

6.1.1 Histograms

We saw earlier how a gamma ray histogram display is used to determine the clean sand response and shale maximum value to calculate the volume of shale log; however, the same type of display can be used to look for differences between wells or individual sand bodies. This is particularly useful when comparing sands with variable feldspar or lithic components as the distribution of each will be different and the difference can be expressed in terms of the average value and standard deviation. Such quantitative differences may help to identify different sediment source areas, depositional environment or proximal–distal relationship of a correlated sand body across a field. When undertaking a multi-well study, the histogram distribution of a property can be used to normalize all the relevant data, providing consistency in the subsequent interpretations.

6.1.2 Scatter plots

Scatter plots can be used to compare logs that represent compatible properties such as sonic and density or data that are unrelated or from different sources, such as laboratory data compared with wireline data. We have seen how the neutron–density cross-plot is used to determine the shale point when matrix and fluid density are fixed, but it is also used to determine lithology using standard graphical overlays (Figure 6.4). In the simplest case of

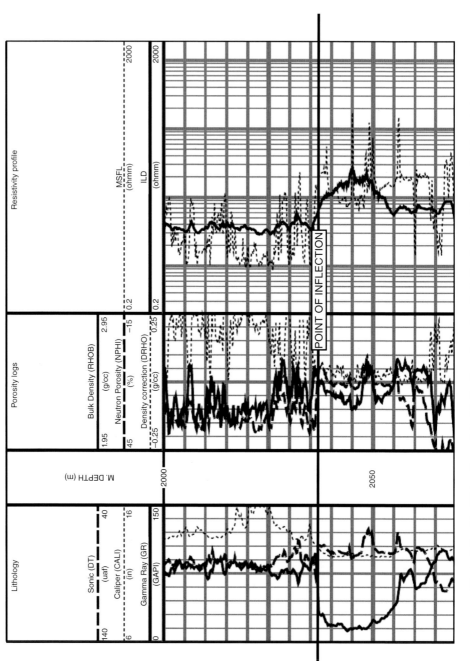

Figure 6.2 Identifying the point of inflection of a suite of wireline logs to determine bed boundaries.

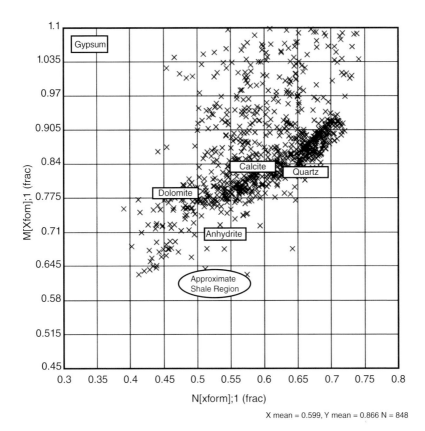

X mean = 0.599, Y mean = 0.866 N = 848

Figure 6.3 Example of M–N lithology plot, where M represents a normalized relationship of the sonic and bulk density and N represents a normalized relationship between bulk density and neutron porosity. This plot is used to identify different minerals and lithologies.

two variables, lithology (matrix density) and porosity and no hydrocarbons, the solution is straightforward and the lithology at every point can be calculated; however, if there is a shale component or more than one porosity type, then new representative end-member values must be defined. This can seldom be done without hard data from cores so that different types of sandstone can be characterized. An example might be where fluvial and aeolian sands are interbedded as in the Triassic Sherwood Sandstone; each sand has different reservoir properties because of the way in which they were deposited and subsequently altered by burial diagenesis, and this is reflected in the porosity, while the matrix density is similar (Bastin et al., 2003). An interactive graphical display of the logged interval and the cross-plot allows individual reservoir sands to identified and highlighted on the plot.

It is often worthwhile plotting cross-plotting logs that have no intrinsic relationship to try to establish some meaningful information (Figure 6.5).

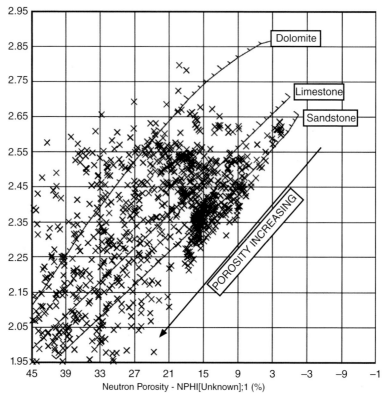

Figure 6.4   Example of a neutron–density cross-plot and interpretation overlay for the appropriate version and supplier of the tool. The cross-plot is used to estimate porosity for a given lithology.

The most common approach is to plot gamma ray against almost any other log property to investigate whether the variation in shale/organic content is related to acoustic impedance from the sonic log or porosity defined by the neutron log, for instance. Shales with high organic content tend to be acoustically 'softer' than shales comprising mainly quartz and clays. Investigating these apparently unrelated properties is sometimes the first step in what ultimately becomes an experiment in multivariate statistics (Doveton, 1994).

Plotting the same type of data from different sources is a standard quality control method in bivariate statistics. A scatter plot of core and log-derived porosity should have a near 1:1 straight-line relationship if depth matching of the two datasets has been done well. If the core porosity data is not over-burden corrected then a constant offset in the data of 1–2 porosity units is to be expected as the log data are at reservoir conditions.

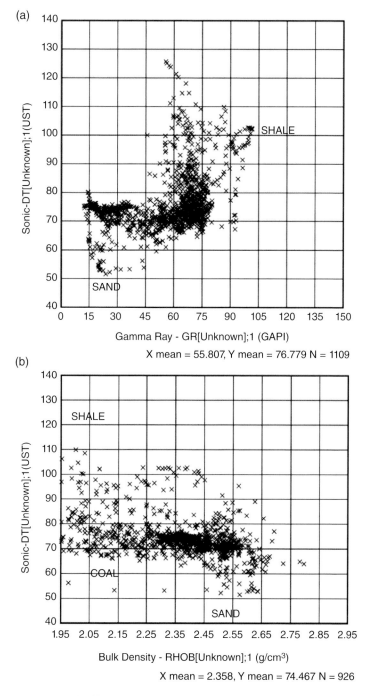

Figure 6.5   Examples of different types of scatter plots used to establish relevant shale points and tool response to different lithologies. (a) Sonic versus gamma ray scatter plot used to estimate sonic response to sand and shale. (b) Comparison of sand and shale response to sonic and bulk density; presence of coal also picked out.

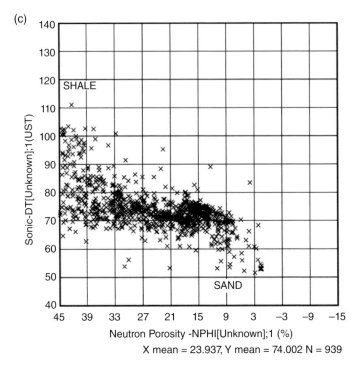

Figure 6.5   (Continued) (c) Scatter plot of raw sonic against neutron porosity used to compare variations to lithology response.

6.1.3   Lithology interpretation

A simple approach to lithology interpretation is to apply a compositional methodology based on the main lithology logs: gamma ray, sonic, density–neutron and PEF. At the most basic level in a clastic sequence we are trying to establish 'good rock' from 'bad rock', high porosity sandstone from well cemented sandstone or shale, and also specific lithologies such as coal, palaeosol and glauconitic sands. Having established a number of lithofacies from core, each perhaps with associated reservoir properties, we want to characterize these in terms of logs. For instance, in a deltaic environment we may recognize the lithofacies and their corresponding log responses shown in Table 6.1.

In this way, each lithofacies has a unique set of log responses and a simple set of equations can be written to identify them in the well. In this case, a unique geological interpretation has also been established.

In a shallow marine environment where the sequence will be sand dominated, it may be necessary to apply a slightly more sophisticated approach to lithofacies recognition using a normalized volume of shale and reservoir

Table 6.1   Typical values of the main lithological determinant logs.

Lithofacies	Gamma ray (API)	Bulk density (g/cm³)	PEF (b/e)
Clean sandstone (channel)	<30	<2.35	<2
Shaly sandstone (overbank)	30–70	<2.4	<2
Cemented sandstone (lag)	<20	>2.55	>2
Shale (floodplain)	>70	<2.45	>3
Coal (swamp)	<30	<2.0	<1

Table 6.2   Discriminant log ranges of shallow marine facies.

Interpreted facies	Volume of shale (V_{sh})	D–N separation (RQI)	PEF (b/e)
Beach/barrier	<0.1	>0.4	>5 (heavy minerals)
Upper shoreface	0.1–0.3	0.2–0.4	<2
Lower shoreface	0.3–0.6	0.1–0.2	<2
Offshore transition	>0.6	0–0.1	>2
Cemented nodules	0	Negative	>2

quality indicator such as density–neutron separation. In this case, the data in Table 6.2 may be appropriate.

Density–neutron separation (RQI) is calculated by normalizing both logs to a range of 0–1 and the subtracting the normalized neutron from the normalized density. These ranges are for a specific shallow marine reservoir and are not applicable globally, and should not be considered as default values. A more comprehensive compositional methodology is discussed in Chapter 8.

6.2   Evaluation of porosity

Having established the lithology and created shale- and hydrocarbon-corrected input logs and defined the corresponding matrix, shale and fluid values for each response, it is possible to move on to the next interpretation step, which is the evaluation of porosity. This is approached in much the same way for each of three main logging tools, although there are specific issues associated with each interpretation. In the following sections the subscripts log, ma and fl refer to the input log, matrix value and fluid value for each tool response, respectively.

6.2.1   Sonic porosity

The interval transit time of a rock is dependent on its lithology and porosity; therefore, a matrix value and fluid transit time must be known to derive a corresponding porosity. Because there is no generic relationship between sonic travel time and porosity, empirical methods must be used. There are two main methods of estimating porosity from the sonic log: the Wyllie time-average equation and the Raymer–Hunt–Gardner equation.

The Wyllie-time average equation in terms of sonic velocity is

$$\frac{1}{V} = \frac{\phi}{V_{fl}} + \frac{1-\phi}{V_{ma}}$$

This can be rewritten in terms of travel time as

$$\phi = \frac{\Delta t_{log} - \Delta t_{ma}}{\Delta t_{fl} - \Delta t_{ma}}$$

where Δt_{log} is the compressional slowness of the formation measured by the sonic tool and the other terms refer to the matrix and fluid properties of the formation. The Wyllie time-average equation can be adapted for unconsolidated sediments by adding an empirical compaction factor $(1/C_p)$ derived from the shale point for an adjacent interval:

$$\phi = \frac{\Delta t_{log} - \Delta t_{ma}}{\Delta t_{fl} - \Delta t_{ma}} \times \frac{1}{C_p}$$

where

$$C_p = \frac{\Delta t_{sh} \times C}{100}$$

The constant C is usually taken as 1 (Hilchie, 1978),

The Wyllie time-average equation was developed for use in well-consolidated sandstones and intergranular limestone or sucrosic dolomites. In the presence of vugs or fractures, the method underestimates the formation porosity; this happens because the tool is reading matrix porosity rather than the total porosity of these more complex reservoirs, including microporosity.

In carbonate rocks, the sonic tool tends to give lower porosity values than the density tool, which is measuring a total porosity including isolated vuggy pores. In gas-bearing sands, the sonic tool will tend to overestimate porosity by a factor of 10–20% because the fluid velocity term is incomplete.

The Raymer–Hunt–Gardner model was designed for poorly consolidated sands:

$$\phi = K \times \frac{\Delta t_{log} - \Delta t_{ma}}{\Delta t_{log}}$$

where K is an empirical constant with values usually between 0.625 and 0.7.

6.2.2 Density porosity

The density tool is considered by most petrophysicists to give the most accurate porosity estimate of the tools available. The equation for calculating the density porosity relates the fluid and matrix values previously determined:

$$\phi_{den} = \frac{\rho_{ma} - \rho_{bulk}}{\rho_{ma} - \rho_{fl}}$$

where ρ_{bulk} is the bulk density of the formation as measured by the density tool; the others terms refer to the fluid and matrix terms. The calculation is most robust where the matrix density is stable, indicating a single lithology or mineral composition. However, because of the wide range of matrix density values for the different reservoir lithologies encountered, it is essential to obtain a robust calibration with the grain density to prevent errors creeping into the estimation. A typical problem can arise where there is a small but significant proportion of a heavy mineral, typically pyrite, distributed through the formation: if not correctly identified, a lower matrix point may be selected, resulting in a higher than actual formation porosity. In complex lithologies such as mixed limestone and anhydrite intervals, it is essential to identify the type and distribution of anhydrite to achieve a satisfactory result.

The density tool generally gives an accurate porosity value in oil-bearing sands; because of the shallow depth of investigation it is only reading the flushed zone where most of the fluid will be mud filtrate. In the presence of gas, however, the results will be less accurate as the fluid is more mobile and will tend to give lower bulk density readings.

6.2.3 Neutron porosity

The neutron porosity is calculated directly from the log response, as the tool is measuring liquid-filled porosity; it is usually calibrated in limestone porosity units and must therefore be corrected for the actual lithology. The relationship between the neutron count rate and porosity can be expressed mathematically as

$$\log_{10} \phi = aN + B$$

where a and B are constants, N is the count rate and ϕ is the true porosity. The constants a and B vary depending on the nature of the formation and require calibration; a limestone and a sandstone will have different log responses even if the porosity is the same. It is essential to know whether the tool has been calibrated for a limestone or sandstone matrix before applying any evaluation technique.

In the presence of gas, the neutron porosity estimated will be less than the true porosity because of the very low hydrogen index of the pore-filling fluid. Corrections can be made, but the best use to be made of the neutron log in a gas-filled reservoir is qualitative. The volume of water trapped within shale leads to higher than expected porosity values affecting the neutron log; the debate about total porosity and effective porosity systems is partly a function of this feature of the neutron tool.

The neutron tool is commonly used in conjunction with the other porosity tools to calculate cross-plot porosity and an associated lithology. In the most common case, the neutron and density log responses are plotted on the x and y axes of a graph, respectively, and an overlay representing three different porous lithologies, sandstone, limestone and dolomite; each line is graduated in porosity units. By plotting each tool response, a dominant lithology and porosity are estimated where they cross; this need not be a unique solution if the lithology is complex.

A commonly used technique to identify reservoir-quality water-filled sands is the separation between the density and neutron logs when displayed side by side; the greater the separation, the better is the reservoir quality. This is especially true when the logs are scaled to emphasize the effect; however, again it is important to know which tool matrix calibration has been applied.

6.2.4   Selection of reservoir porosity

There are three stages in the selection of the optimum reservoir porosity:

- comparison of core and log porosity data
- comparison of different log-derived porosity results
- assignment of final porosity on a level-by-level basis.

Comparison of core and log porosity data is best done using a simple histogram plot of the distribution of each of porosity to be compared. Check the upper limit of porosity for each interpretation zone and set this to be the maximum allowable for that zone. Compare the histogram of core against log porosity and ensure that the descriptive statistics (mean, standard deviation, etc.) are comparable and then select the log porosity that best represents the core data in each zone. It is important to ensure that the comparison takes place where borehole conditions are optimum and rugosity is minimized.

Having identified which log porosity gives the best agreement with core data where borehole conditions are good, it is necessary to compare the calculated porosity of each tool where hole conditions are poor; density and neutron tools perform poorly in badly washed-out sections. The sonic log is less affected by borehole conditions but the limitations of the time-average equation and lack of a compaction correction might have a greater negative effect than borehole conditions. It is recognized that washouts tend to result in an overestimation of porosity; therefore, it is good practice to select the lowest calculated porosity in these intervals.

The other borehole problem to be aware of is tool sticking as indicated by the tension curve. Tool sticking tends to affect the neutron and density logs most because of the tool construction (pad and calliper device) and results in lower calculated porosity values. It is recommended that the sonic porosity be used where tool sticking is recognized.

The assignment of final porosity values requires that the definitive calculated porosity from each hole section is merged into an optimum curve: the log porosity that most closely compares to the core data with the minimum porosity where the borehole is washed out and the sonic where the tools are seen to be sticking. The results should also be truncated where they exceed the previously determined zone maximum. In wells with other lithologies such as coal or volcanics in which the standard porosity evaluation does not apply, the results of the calculations should be set to zero or a null value.

6.2.5 Total and effective systems

Whether the porosity determined should represent the total porosity or the effective porosity has been discussed long and hard for several decades and continues to be a subject of much debate among petrophysicists. Total porosity is that generally measured from core analysis and should represent the maximum porosity calculated by any other method; effective porosity discounts that volume filled by clay-bound and irreducible water and is therefore always less than or equal to total porosity depending on the volume of shale (Figure 6.6). In the model presented, the porosity measured by the neutron log includes the volume of clay minerals containing hydrogen as part of the crystal structure, not water.

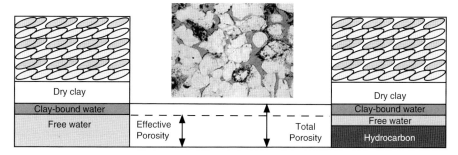

Figure 6.6 Log analysis gives total porosity including that associated with clay-bound water. Core analysis may also give total porosity depending on the cleaning and drying methods applied. For volumetric calculations we need effective properties.

Different operating companies have their preferred methods of interpretation and will usually stick with one system or another. It is important that at the start of a project the approach is agreed and the subsequent interpretation procedures adhere to one system or another. Most log analysis software packages have the ability to calculate porosity and water saturation in either system. Increasingly, because of the ease and speed of modern computational systems, both total and effective porosity will be calculated. In 3D reservoir models, the overburden-corrected, effective porosity should be propagated through the cells.

6.3   Evaluation of water resistivity

An estimation of formation water resistivity, R_w, is required for the calculation of water saturation, S_w. If an uncontaminated water sample can be obtained from a well through a drill stem test or a wireline formation test, this is the most reliable method of determining R_w. However, a single sample is seldom sufficient to characterize the formation water throughout the field as both geographical and vertical variations can be expected. Be aware that the formation salinity in the aquifer may be different to that of the hydrocarbon-bearing reservoir due to the hydrocarbon filling history of the trap.

Log-derived estimates of R_w are often used to extend the dataset or to look for marked variations. The process of evaluation tends to be iterative in that as more data are gathered a better evaluation is developed. R_w is dependent on two linked variables, temperature and salinity: as these properties vary in the reservoir, so does R_w. There are a number of methods that can be used to determine R_w and these should be compared to find the most consistent for a particular field or zone.

6.3.1   SP method

The spontaneous potential (SP) method can be used to determine R_w where there are strong salinity contrasts between the mud filtrate and the formation water, in thick, clean sandstones with well-defined shale packages. Ideally, this method should be applied in sands below the oil–water contact and will only work where a water-based drilling fluid has been used. If the raw SP curve is featureless, then one or more of these conditions has been infringed.

The first step in this method is to check the curve for any shift in the baseline over the evaluation interval; look for systematic variations in the tool response in thick shale packages. If there is no observable shift in the baseline then the interval can be treated as one interpretation interval, otherwise zones of constancy will need to identified and treated separately. A scaled SP log should be created over each interpretation interval by setting the shale baseline to zero and adding or subtracting this response from the raw curve. Once created, the scaled log will deal with the problem of baseline shift and all intervals can be treated as one; this is also known as the static self potential (SSP).

Other information required for the evaluation are the formation temperature and the resistivity of the drilling mud (R_m) and the mud filtrate (R_{mf}), which can be obtained from the log header. The resistivity values are corrected to the relevant bottom hole temperature (BHT in °F) on a level-by-level basis and used to determine the equivalent formation water resistivity, R_{we} from the equation

$$R_{we} = R_{mf} \times 10^{SP(61+0.133BHT)}$$

The value of R_{we} is then corrected for the variations in salinity and formation temperature. Most log analysis software tools will have this methodology built into their workflows, so it up to the interpreter to ensure that the input data are validated and the results are meaningful in the context of the reservoir description. This process should be repeated for all wells drilled with water-based mud and the results compared in an effort to identify a field-wide value.

6.3.2   Resistivity cross-plot method

This procedure is applicable in wells drilled with water-based mud and where there is an identifiable water zone. The method is based on the Archie definition of formation factor, F, which can be written as

$$F = \frac{R_o}{R_w} = \frac{R_{xo}}{R_{mf}}$$

where R_o is the resistivity of a rock fully saturated with formation water, R_{xo} is the resistivity of a rock fully saturated with mud filtrate and R_{mf} is the resistivity of the mud filtrate.

A linear cross-plot of R_o (deep resistivity) against R_{xo} (microspherically focused resistivity) for a clean water-saturated rock will produce a line with a slope of gradient R_{mf}/R_w, from which R_w can be calculated when R_{mf} is known (Figure 6.7). The challenge with this method is finding hydrocarbon-free intervals to use, as any trapped oil will have the effect of increasing R_o relative to R_{xo}, i.e. points that would have been on the water line are shifted down and to the right of the plot. In this case, fitting the line is achieved by using the points in the opposite quadrant and drawing the line through the origin. Where there are a limited number of points, it is possible to relax the V_{sh} cut-off in an attempt to include more data; however, as either the mud filtrate or the formation water are likely to be relatively fresh (<20,000 ppm NaCl), this

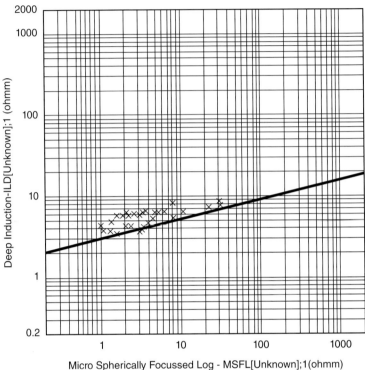

Micro Spherically Focussed Log - MSFL[Unknown];1(ohmm)

X = t(Y); LD = 10"(0.191135") (log)MSFL - 0.627865) CC: 0.7063 N = 32

Figure 6.7   Example of a cross-plot of deep and shallow reading resistivity tools used to establish the formation water resistivity when the resistivity of the mud filtrate is known. The slope of the line gives R_w.

seldom makes a large difference. Having drawn a line and determined the gradient, investigate the distribution of the viable points; if these come from the same interval in the well, note the depth and convert the value of R_{mf} from the log header to a value that corresponds to the temperature at that depth. Calculate the R_w from this value of R_{mf} and the gradient; R_w will then already be corrected for the relevant temperature. Where the valid points are scattered through a larger interval, take the mid-point to calculate the temperature-corrected R_{mf}.

The same procedure should be applied to all the wells drilled with water-based mud and logged with a microspherically focused resistivity tool. The individual estimations of R_w must be corrected to a common depth/temperature value for use in a field-wide solution. This method can produce wildly variable and ambiguous results and should be used with caution; however, the results should be compared with other solutions for consistency.

6.3.3   Pickett plot

The Pickett plot is the most commonly used method of R_w determination in the absence of formation water samples. The method can be applied regardless of the type of drilling mud, provided that there is a water-bearing, clean sand interval; it cannot be applied, however, until porosity has been determined. The Pickett method simply uses a logarithmic version of the Archie equation:

$$\log R_o = \log(aR_w) - m\log\phi$$

where a and m are the Archie coefficient and porosity exponent, respectively.

A bi-logarithmic cross-plot of R_o and ϕ should yield a line with a slope of $-1/m$; this is the so-called 'water line'. In the case of zero porosity ($\phi = 0$), R_o is equal to aR_w, allowing R_w to be determined from the intercept of the straight line with the line equal to $1 = a$. As with the previous method, any hydrocarbons in the interval will have the effect of increasing R_o and shifting points to the right of the line, so that the water line is usually positioned to the left edge of the data cluster (Figure 6.8). The results must be temperature corrected to obtain a value of R_w usable in subsequent field-wide analysis.

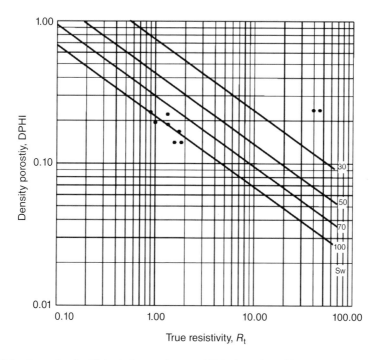

Figure 6.8   Example of a Pickett plot used to establish the value of formation water resistivity when the Archie parameters a, m and n are known.

The Pickett plot can be used in a number of ways, but to determine R_w it is necessary to have robust values of a and m from core analysis. If a reliable water sample has been collected, the plot can be used to determine a and m from a log-derived porosity. Most log analysis software packages have built-in Pickett plot routines.

6.3.4   Apparent R_w method (R_{wa})

Like the preceding methods, this procedure relies on another form of the Archie equation applied to clean water-bearing sandstone:

$$R_w = \frac{R_o}{F}$$

which thus becomes

$$R_{wa} = \frac{R_t \times \phi^m}{a}$$

because R_o would be water bearing.

In hydrocarbon-bearing sands, the value of R_o will be greater for a given formation factor, hence the computed R_w will be overestimated and is termed the apparent formation water resistivity, R_{wa}. If this computation is made over an interval containing hydrocarbons as well as fully water-saturated sandstones, then the minimum value of R_{wa} is equivalent to R_w. It is usually possible to correct the estimation for clay-bearing sandstone, by introducing a term for R_{sh}. It is essential that the Archie exponents a, m and n have been accurately measured in the laboratory using a sufficiently saline brine that any shale effects are negated; if there is any doubt, use the default values. Selection of a meaningful value for $R_{wa} = R_w$ is largely subjective and dependent on numerous external artefacts, including variable salinity, temperature and estimation in low-porosity rocks.

6.4   Estimation of water saturation

Determining the water saturation is often the final step in a log analysis exercise (Figure 6.9). However, it is also important to find out whether a reservoir contains moveable hydrocarbons and whether they can be produced water free. There are two cases to consider, that of the clean, clay-free sand and that of the clay-rich sand: we have already discussed the determination of shale or clay content. It is often wise to assume that the reservoir contains clay and shale, because in the rare case where it does not the algorithms default to the clean sand case and the simple Archie equation is sufficient.

Calculation of Water Saturation

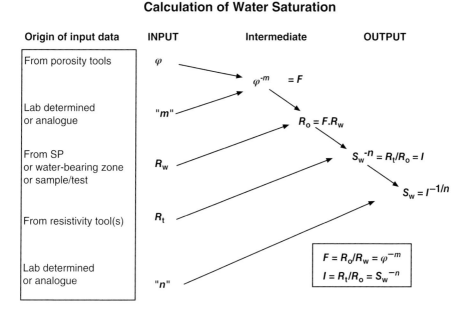

Figure 6.9   Process diagram used to calculate water saturation using the Archie equation for clean sands.

6.4.1   Clean sands

The water saturation of clean sands can be determined using the combined Archie equation:

$$S_w = \left(\frac{aR_w}{R_t \phi^m} \right)^{\frac{1}{n}}$$

where a is the Archie porosity coefficient, m is the cementation exponent and n is the saturation exponent; all three values should be determined from core analysis. R_w is the formation water resistivity determined from an uncontaminated water sample, R_t is the response of the deep investigation resistivity tool appropriately corrected for the borehole environment and ϕ is the final porosity previously determined. All log analysis packages will have this basic calculation built into the workflow.

By substituting R_{mf} for R_w and R_{xo} for R_t, it is possible to estimate the water saturation in the flushed zone, S_{xo}. By comparing S_w with S_{xo}, it is possible to determine whether the hydrocarbons are moveable; if $S_{xo} = S_w$ then no fluids were moved and the formation is either tight or plugged with fines. The ratio S_w/S_{xo} is known as the moveable hydrocarbon index (MHI). A further extension of this procedure allows the interpreter to determine S_w without needing porosity or a and m; this is known as the *ratio method*.

6.4.2   Shaly sands

The Archie equation presupposes that the rock framework is not electrically conductive, in other words, a perfect insulator. In reality, the generally ubiquitous presence of clay minerals in sandstones adds a conductive element that causes the Archie equation to overestimate water saturation. As has been discussed previously, there are two types of 'shaly sands', those with detrital shale layers, clasts or grains dispersed by bioturbation and sands that contain authigenic clays as a result of diagenesis. The commercial impact of finding a universal solution to the resistivity equation has kept industrial petrophysicists engaged for most of the last 40 years. The general shaly sand equation takes the form:

$$\frac{1}{R_t} = \frac{S_w^{\,2}}{FR_w} + X$$

where X is the conductivity of the shale component.

Worthington (1985) grouped these various solutions into two families: those that considered shale as a homogeneous conductive component where the equation depends on an accurate estimate of V_{sh}, and those that viewed clay as a separate conductive ionic layer around sand grains; in effect, the conductivity of the clay component is a function of the cation-exchange capacity (CEC) of the various clay types present. In the end, all these equations default to the Archie equation in the absence of clay or the presence of high-saline brines.

The Simandoux equation (Simandoux, 1963) is probably the best known of the V_{sh} solutions; the modification is in the calculation of X where $(1 - V_{sh})$ is used.

$$\frac{1}{R_t} = \frac{S_w^{\,2}}{FR_w} + \frac{\varepsilon V_{sh}}{R_{sh}}$$

where ε varies with S_w such that if $\varepsilon = 1$ then $S_w = 1$, and if $\varepsilon < 1$ then $S_w < 1$.

The 'classic' shaly sand equations sometimes provide unreliable water saturation estimations. Intrinsic weaknesses include the use of shale resistivity from shale interbeds, where clay mineral species and morphology may differ drastically from those in porous reservoir zones and shale indicators that estimate volumes rather than active surface areas.

Waxman and Smits (1968) provided an example of the second type of shaly sand solution; this requires estimates of the concentration of exchange cations and water saturation in addition to values of depth, shale volume, porosity and resistivity. Realistic shaly sand evaluations depend upon knowledge of clay mineral species, surface areas and cation-exchange capacities. Most of these input data are determined in a laboratory and require core data.

The basic equation for the Waxman–Smits model is

$$\frac{1}{R_t} = \frac{S_w^{\ 2}}{F^* R_w} + \frac{B Q_v S_w}{F^*}$$

where B is the specific counter-ion activity and was calculated by Waxman and Smits (1968) using

$$Q_1 = \frac{-1.28 + 0.225 \times ZT - 0.0004059 \times ZT \times ZT}{1 + R_w^{\ 1.23} \times (0.045 \times ZT - 0.27)}$$

where ZT is a temperature gradient correction. Qv is the cation-exchange concentration determined from laboratory experiment.

Weaknesses in the Waxman–Smits approach led to the development of the dual water model (Clavier et al., 1977) that considers two kinds of water in a shaly formation: bound water and free water (Figure 6.10). The bound water adheres to the shale surface as a thin layer and cannot be produced. Free water is all other water including any irreducible water: not all 'free' water is producible. Total porosity equals bound and free water plus hydrocarbons.

The concept behind the model is that the charged ion concentration in the water bound to the shale surfaces is quite different from the free water.

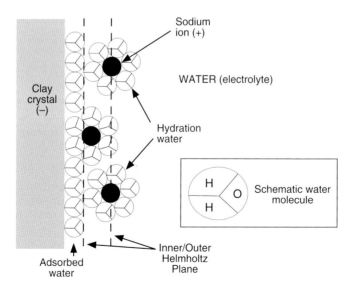

Figure 6.10   Model of clay bound water and the distribution of exchangeable cations on a clay surface; the greater the number of exchangeable cations, the greater the CEC and the greater the surface conductance of the clay.

Consequently, the conductivity of the bound water is different. This occurs because the large surface area of negative charge presented by the shale surfaces attracts the positive end of the dipolar water molecules. This leaves the negative end of the water molecules unpaired: anions are attracted to them. Thus the outside surface of the bound water layer has a tendency to be rich in sodium ions, anions, in preference to chloride ions, cations. This model uses the following equation:

$$\frac{1}{R_t} = \frac{S_w^{\,2}}{F^* R_w} + \frac{(C_{bw} - C_w) V_Q Q_v S_w}{F_o}$$

Shaly sand analysis is a complex and challenging interpretation process. It is necessary to know many different intrinsic shale properties of the reservoir, properties that are difficult to determine and are therefore seldom collected routinely. The interpreter is thus reduced to making assumptions that are often unsubstantiated. The flowing 'rules of thumb' may help to improve an interpretation:

- Using adjacent shales to determine the R_{sh} value may not always be appropriate; check with the geologist for alternatives or use an equation that does not require a value for R_{sh}.
- Kaolinite and chlorite tend to have extremely low CEC values, whereas illite and smectite have high CEC values.
- Where the formation water salinity is greater than about 20,000 ppm NaCl, the effects of authigenic clays are limited, although the volume impact of detrital shale must be considered.

6.5   Summary

The estimation of porosity and water saturation is the end product of a petrophysical study and the importance of the results in volumetric estimation cannot be emphasized enough. The uncertainty in each measurement and the different scales at which they are applied must always be taken into consideration when making the calculation. Sometimes it may be sufficient to say that effective porosity may vary by 2 porosity units. about the zone average, but what does the zone average represent?

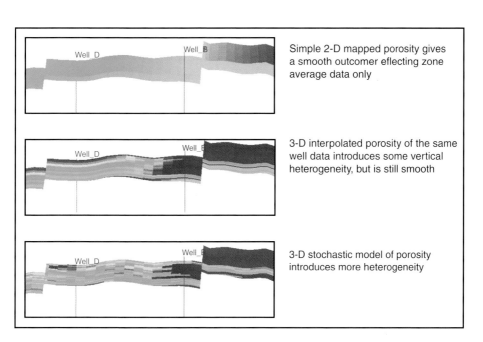

Figure 10.2 Porosity distribution: mapped, interpolated and stochastically distributed showing the increasing heterogeneity in the property. Source: Reproduced courtesy of Roxar Limited.

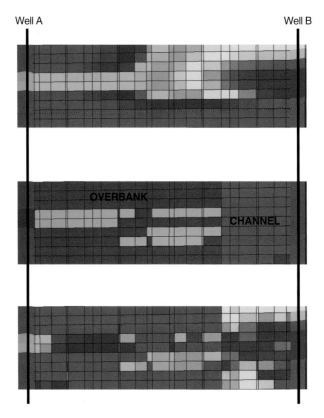

Figure 10.3 Facies-constrained porosity distribution. (a) The interpolated porosity model honours the well data but results in a smooth distribution between the wells; (b, c) a simple threefold scheme of channel, overbank and floodplain facies allows the porosity seen in the wells to be distributed meaningfully, capturing the rapid changes laterally in the model. Source: Reproduced courtesy of Roxar Limited.

7

Petrophysical Workflows

This chapter describes two typical workflows for a petrophysical study; the first is a 'quick-look' analysis facilitated usually by a default solution in a log analysis programme and the second is a more complete approach applicable for a detailed post-well study or a field study where more data are available. The quick-look approach can also be used at the wellsite with the help of a spreadsheet, calculator or the back of an envelope! Appendix 1 has a worked example of a single-well petrophysical study from database to report that can be used as a template.

7.1 Data management

Standard wireline and LWD measurements are made typically every 6 in (0.152 m), while high-resolution tools measure data at an even finer scale. Given an average well measuring say 3000 m (10,000 ft) along hole, we might have 20,000 measured values for a gamma ray to file and store. Even over the reservoir section (~200 m) in an average well there may be 100,000 data points acquired, which will need to be carefully checked for quality of the measurement and, most importantly, the depth.

The initial (raw) data provided by the logging contractor will be presented as a series of log runs associated with a particular hole size: where intermediate logs are run it would be prudent to re-log the complete hole section again, just in case borehole conditions or the tools have changed between runs. Intermediate logging runs are made before the well section total depth is reached, usually to safeguard information from a reservoir before drilling on. Working with the logging engineers and wellsite geologist or petroleum engineer, it should be possible to piece together the history of each logging job on a well, but this should be done as soon as possible after the event as memory has a tendency to fade in busy operational conditions.

Petrophysics: A Practical Guide, First Edition. Steve Cannon.
© 2016 John Wiley & Sons, Ltd. Published 2016 by John Wiley & Sons, Ltd.

The initial data will have only minimal processing before being interpreted and, although this 'quick-look' analysis is valuable, it will normally be superseded by a complete interpretation, often incorporating core analysis data. All of these interpretations are part of the 'history' of the well evaluation and should often be treated as work in progress, as the results will be reviewed and reworked time and again as new data or new wells become available. At each stage, a definitive well database should be constructed with a readily understood naming convention for input data and results. Many companies have their own log naming conventions and database constructions and it behoves the new petrophysicist to adopt a method that has probably been developed over many years.

The data are commonly stored in a mainframe computer using some bespoke database, provided by either the logging service company or an international software company. There are many examples to choose from, but the content of these large databases is seldom complete in all respects. Often the data are loaded under a well name only, without many of the other cultural data to pinpoint the location; it is common for the data to be loaded without the borehole environmental data necessary for even a simple log interpretation. It is often very beneficial to go back to the basic mud logging and drilling reports to fill in the well history at the same time as gathering the data needed for an interpretation. My preference is always to build a well or project database from the final set of quality-controlled logs provided by the service company; it will still be necessary to splice and possibly depth shift the logs to create a full well computer-processed interpretation (CPI).

Most modern log analysis packages offer a number of predefined workflows, including a quick-look routine and a step-by-step logic that follows the order of analysis described previously; from lithology and shale estimation through porosity and water saturation calculation. The output format is also fairly standard, generating a default CPI. Fortunately, the software tools also allow the experienced user to define their own algorithms and output formats, consistent with company templates. There is, of course, a simpler, older methodology still valid when a wellsite or data-room log evaluation is required and there is no powerful software solution available, just a calculator or hopefully a spreadsheet!

7.2   Quick-look interpretation

With a quick-look interpretation, the objective is to establish the presence of hydrocarbon-bearing, reservoir-quality formations that may be worthy of production testing. In this case we are using primarily the basic log data, hopefully supported by some offset well data to provide a local context: in a wildcat well we may not have any support data and an interpretation becomes more intuitive.

If you are at the wellsite to witness the logging runs at total depth (TD) or perhaps to analyse the logs, what relevant data may you have to hand? – a mudlog with cuttings lithology, gas readings and rate of penetration data; drilling mud information such as mud weight and resistivity measured at the surface; mud filtrate resistivity; mud temperature at the surface with some idea of the temperature gradient; and hopefully that offset information, such as formation water salinity and values for the Archie parameters. Even if you are onshore waiting for the digital data to come in by e-mail, you should have this information to hand at all times, because logging, like coring, always happens at night!

However the data arrives, be it by a data-link or e-mail (or even fax), the logs should be displayed and checked for obvious discrepancies such as depth shifts or zones where the tool string may have been stuck. Check that the logger's TD and the casing shoe depths are consistent with the latest drilling reports and that the correct datum is being used. Hole rugosity can be evaluated with the calliper log, having first checked that the tool is reading correctly in the casing string. The sonic tool can also be checked for accuracy of the reading; the sonic transit time in the casing should read ~47 µs/ft. The density correction curve should also not exceed ~0.2 g/cm^3, except where the borehole is clearly washed out. Finally, the resistivity curves should also tell a story; in an oil-based system, the shallow reading curves will read higher than the deep reading tools, and the opposite in water-based mud, provided that the resistivity of the mud filtrate is less than the resistivity of the formation water (another reason why offset data can be useful).

The next step is to scan the logs to identify potential reservoir zones, be they clastic or carbonate. First look for non-shaly, clean intervals using the gamma ray as a guide. In thick homogeneous clastic rocks the resistivity is the best indicator of the presence of hydrocarbons; where it is low it will be water bearing and where readings are high it may be gas or oil bearing. The porosity logs should be reviewed next and where it is high in conjunction with a positive resistivity response you have a zone of interest. In carbonates, because of the variability of rock types, a porosity log is better at identifying potential reservoirs and where the resistivity is correspondingly high there is a zone of potential interest. This routine is very simplistic because feldspathic or lithic-rich sands may have a high gamma response and could be considered to be shale. Thin-bedded sands or low resistivity pay can easily be overlooked without a more robust interpretation routine.

A computer based quick-look interpretation will require clean sand and shale values for the gamma ray, and also matrix and fluid density data to calculate the porosity from the density log. The neutron–density may also be selected to estimate porosity and used to compare with the density–porosity; these will both be total porosity calculations. Because this is a quick-look interpretation, applying the Archie equation to estimate water saturation is sufficient; using the default values for m and n is also acceptable unless there are calibrated offset data to hand. As the Archie parameters are defined, the only remaining variable in the equation is formation water resistivity. A Pickett plot of porosity against formation resistivity in a water zone should be constructed to estimate R_w, having fixed the slope of the line. If no water-bearing reservoir interval has been penetrated, R_w should be estimated from a local or regional water sample. Comparison of a Pickett plot-derived R_w and the selected regional value may show differences. This could be for a number of reasons, including the presence of minor hydrocarbons, inaccurate porosity estimation and different formation water chemistry.

The final step is to present the results of the analysis in both graphical and tabular form. Usually, in a single-well analysis, reservoir zones are simple stratigraphic intervals agreed with the geologist and a distinction between different fluids where appropriate: more complexity is seldom required until the development stage. Average values for porosity, water saturation and net:gross are usually reported; permeability ranges may be included, but choosing to average permeability at this stage may be counter-productive. Both porosity and saturation should be averaged as a function of the net thickness of the interval under investigation. Whether to apply some form of porosity or saturation cut-off at this stage may also be premature, but is often *de rigueur.*

Graphically, the quick-look CPI should show the input logs and also the calculated logs in sequence from left to right (Figure 7.1). The final track will normally be the summation of results showing dominant lithology, porosity and hydrocarbons and both moveable and irreducible water saturation. A plot of porosity multiplied by hydrocarbon saturation $(1 - S_w)$ displayed in the calculated porosity track is a useful way to see prominent hydrocarbon-bearing intervals.

The quick-look interpretation will be revised and updated more than once as more data from the subject well and future wells become available. It is very much a preliminary result or should be considered as ' work in progress'.

7.3   Full petrophysical interpretation

When there are more data available, usually after the well has been abandoned or suspended, the petrophysicist has more time to review all the results from core analysis, fluid analysis and potentially well test data. If the well is part of an appraisal and development programme, there may be a significant review element included in the work programme. This may include the application of rock physics analysis to support seismic attribute models and wider lithology and fluid prediction workflows aimed at better prediction of hydrocarbon and reservoir distribution.

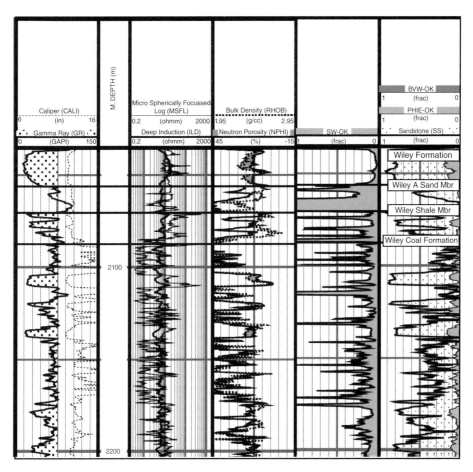

Figure 7.1 Simple computer-processed interpretation (CPI) of a suite of wireline logs: primary input logs in columns 1–4 and S_w estimation in column 5 with the distribution of shale, matrix, porosity and BVW in column 6.

A visit to the core store or core laboratory can be an enlightening experience for the petrophysicist, especially when accompanied by the reservoir geologist; someone who can point out the variability in the rocks and how they may be reflected in the log data. In the fortunate situation where the whole reservoir has been cored, it is possible to calibrate most of the log interpretation outputs by accurately matching the two datasets. The first step is to make any depth shifts required to match to the two data sets; comparing the wireline gamma with core gamma measurements is one approach. Alternatively, a sometimes a more robust way is to compare core plug grain density measurements with the bulk density log. Depth shifts are to be expected, especially where a series of cores are cut sequentially, as core recovery is seldom 100% and a degree of shifting and stretching of the material is required.

Visual inspection and a detailed description of the core will identify such features as cemented layers that appear as tight streaks on a porosity log; gamma ray spikes within otherwise homogeneous sands can be identified as shale interbeds in a fluvial channel or heavy mineral lags associated with beach deposits; hydrocarbon staining or live oil seen in the core will aid accurate oil–water contact identification. It is now when the petrophysicist should be able to appreciate that a single set of input parameters may not give the best outcome. This is especially true when deriving porosity–permeability relationships and variations in the Archie parameters. It is also important that representative samples are selected for any petrographic, clay mineralogy or special core analysis.

Using core analysis data to calibrate the preliminary log interpretation, it is necessary to have porosity and permeability measurements made at overburden conditions to compensate for pore compressibility. The correction factor for poorly consolidated sediments can be significant, up to 0.85, but reliable measurements are difficult to make. Lithified sandstones with high effective porosity will require an overburden correction factor of 0.95–0.98, whereas lower porosity, well lithified or cemented sandstone will require little correction. It is common to plot core-uncorrected core porosity measurements against log estimation to evaluate the need for overburden corrections, as only a subset of samples will be selected for the test; they are costly and can be time consuming.

Routine core analysis also generates a grain density measurement for each plug. These data can be used to quality control the other measurements as values from sandstones less than about 2.63 g/cm^3 should be queried; they may not have been sufficiently well cleaned or dried. Samples heavier than about 2.69 g/cm^3 may contain heavy minerals or carbonate cements; limestone should be in the range 2.71–2.75 g/cm^3. Plotting the grain density as a histogram will provide a representative value for matrix density when used in calculating porosity from the density log.

Plotting the overburden-corrected core porosity against the log-derived porosity will also provide information on the reservoir fluid density (Figure 7.2). With the log porosity on the x-axis and core porosity on the y-axis, where core porosity is zero should be equivalent to the grain density. A straight line from this point to unity on the y-axis will indicate the apparent fluid density. This procedure must be carried out separately for each hydrocarbon and water leg. The results should be similar in value to the parameters used in the quick-look analysis and can be used in the calculation of the definitive porosity. Where the data are consistent in a number of wells, these values should be used in field-wide interpretation.

The estimation of water saturation can now be updated if SCAL measurements of m and n are available. The cementation exponent (m) should be

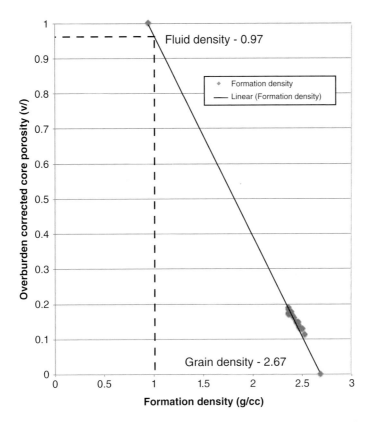

Figure 7.2 Cross-plot showing overburden-corrected porosity against formation density from wireline data, used to determine fluid density and matrix or grain density.

between 1.8 and 2.2; however, if there is a large range of values it may indicate the presence of highly variable rock types, especially in carbonates. By plotting the logarithm of formation factor (F) against porosity, according to Archie:

$$\log F = -m\log\phi$$

The gradient of the line gives the value of m. and the higher the value of m the greater is the calculated water saturation for a given porosity.

The saturation exponent n can be estimated by plotting the logarithm of resistivity index (I) against logarithm of water saturation (S_w), and again from Archie:

$$\log I = -n\log S_w$$

The gradient of the line gives the value for n, and the higher the value of n the greater is the calculated water saturation.

With both m and n fixed, there is little flexibility in choosing a value for formation water resistivity to achieve a water saturation of unity (100%) in a proven water-bearing reservoir. Where an uncontaminated water sample has been recovered and robust water salinity established, a choice has to be made between honouring m or R_w; fortunately, there is always reasonable doubt in the quality of the porosity estimate to leave a little room for manoeuvre. Reviewing the scatter in the data and revising the value of m can usually resolve the problem. If the core porosity data come from the hydrocarbon leg, it is possible that diagenesis in the water leg has continued post-migration and reduced porosity, making the data unrepresentative. When all else fails, use a saturation height relationship if you can confidently establish a free water level; in all cases, it is valuable to compare both sets of results.

7.3.1   Permeability estimation

An initial porosity–permeability relationship can also be established using the routine core analysis data (Figure 7.3). The porosity data should be overburden corrected where possible and calibrated with liquid permeability results if possible. Plotting porosity on the x-axis and the logarithm of permeability on the y-axis gives a y-on-x straight-line regression relationship of the form

$$k = 10^{(a+b+\phi)}$$

where a and b are constants for each facies or reservoir zone. This simple approach usually underestimates the lower end the permeability range and overestimates the higher range; using a power relationship can correct for this inherent deficiency of a straight-line regression. As the graphing usually takes place in a spreadsheet or petrophysical package, it is possible to analyse the relationship for different intervals in addition to facies.

Relationships can be established on a field-wide basis or individual reservoir zone, facies or rock type over different porosity and permeability ranges. If a single empirical relationship holds for groups of facies, then these can be combined if there is no inherent geological difference between them; sheet sands should not be combined with channel sands even if there properties are apparently the same because they will have different depositional characteristics that may control flow in the reservoir. It may be necessary to truncate the upper limit of predicted permeability so that they do not exceed the maximum matrix or intergranular permeability measured on core for a particular facies or zone. This manipulation of results is often necessary to prevent the predictions from becoming too extreme.

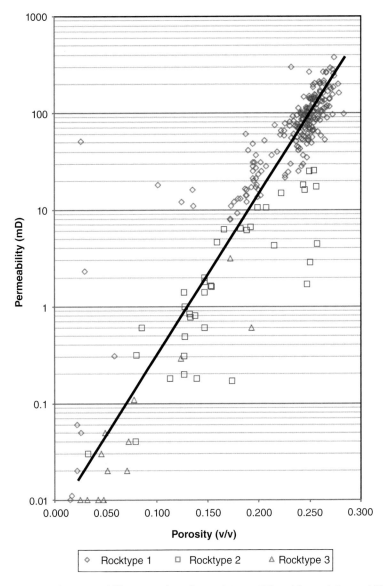

Figure 7.3 Porosity–permeability cross-plot of core data partitioned by rock types. Although a single straight-line relationship is shown, each rock type may have its own predictive relationship.

7.3.2 Evaluation of cut-off parameters

The determination of cut-off parameters has the goal of excluding those parts of the reservoir that does not contribute to either volume or production, or both, of hydrocarbons. The particular contribution will depend on whether the objective is to determine parameters for STOIIP calculation or for a dynamic model of the reservoir; for the former we are evaluating volume of shale, porosity and water saturation and for the latter the contribution of permeability. Cut-offs should be investigated by facies and zone on a well-by-well basis; however, there is always a limiting conditioning that should guide the procedure, namely sufficiency of data for each of these classifications. It is always worthwhile starting with the smallest sustainable volume of data and grouping these where appropriate if the cut-offs are similar.

The selection of cut-offs is a very subjective exercise and can be highly contentious, especially in equity discussions, where a difference of a single porosity unit can have a large impact on in-place volumes. If the selection and application of a robust facies classification have been successful, the non-reservoir intervals in a well may be automatically eliminated; however, it is more likely that a V_{sh} cut-off will be required to determine net sand. Application of a porosity cut-off will limit the non-reservoir intervals and an S_w cut-off will reveal the net pay; what these values should be will often be found after a lengthy iterative process involving the geologist and engineer and also the petrophysicist.

There are some simple rules of thumb that might be used as a starting point for the evaluation of cut-offs:

- A V_{sh} cut-off of 0.5 is often a reasonable starting point to exclude shale-dominant intervals; this can be extended or reduced as required to determine net sand.
- A porosity cut-off can often be found from the porosity–permeability x-plot; the porosity equivalent to 1 mD will work in oil-bearing reservoirs and 0.1 mD for gas zones as net reservoir.
- Water saturation equivalent to 0.70 is often equivalent to initial water-free production and can be used as the net pay identifier.

Ringrose and Bentley (2014) advocate the use of a total property approach to modelling, whereby no cut-offs are applied; rather, a facies or rock-type classification can be used to exclude non-reservoir (see Chapter 10 for further discussion).

7.3.3   Determination of zone averages

For the purposes of property mapping, it is generally necessary to calculate facies averages and property averages for each reservoir zone, particular V_{sh} and porosity. The distribution of average values for each facies and its properties can then be examined to see whether trend mapping is needed to aid the distribution of properties in the inter-well areas and volumes.

Porosity can be averaged over a given interval as a simple arithmetic average or as the mid-point on a cumulative distribution plot for greater accuracy, especially if the interval is particularly heterogeneous and sampling is biased to the better sands. This is simple to do on a spreadsheet allowing a graphical and numerical comparison of the results (Figure 7.4). When calculating an average net:gross value for a number of zones, it is important to remember that it is not sufficient to just add the values, but that the thickness of each interval must considered in the calculation.

If STOIIP is to be calculated by inter-well correlation and mapping, it is desirable to calculate porosity-weighted average values of S_w for individual facies occurring above the free water level; all other static properties will be unweighted arithmetic average values. Averaging permeability should be done with the results of any well-test data taken into consideration; use of permeability–thickness calculations is often revealing on an individual well

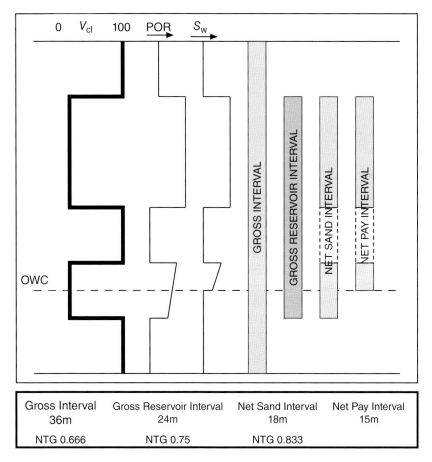

Gross Interval	Gross Reservoir Interval	Net Sand Interval	Net Pay Interval
36m	24m	18m	15m
NTG 0.666	NTG 0.75	NTG 0.833	

Figure 7.4   Net-to-gross calculation using V_{cl}, porosity and S_w as potential cut-offs to distinguish net reservoir, net sand and net pay.

basis. Failing this level of detail, a geometric mean permeability average will usually provide an initial set of input values.

If stochastic methods are to be used for the distribution of properties, S_w values should be assigned to each facies with reference to the height above the free water level previously determined.

7.3.4   Reporting

Presentation of results is very similar to the quick-look analysis with graphical and tabular formats; however, to capture the complete workflow, especially with a multi-well study or field-wide study, a 'proper' report is required. It is essential to document the individual steps in the workflow and to record input data, processing, analysis, interpretation and outcomes. Even interpretation approaches that were tried but not used should be recorded as it may help future workers avoid blind alleys.

A typical report will have the following sections:

1. Objectives
2. Reservoir description
3. Data preparation
4. Log analysis
    4.1 Shale volume estimation
    4.2 Archie parameters
    4.3 Formation water resistivity
    4.4 Matrix density
    4.5 Porosity estimation
    4.6 Saturation estimation
    4.7 Permeability estimation
5. Rock typing
6. Conclusions and recommendations

Each section will also contain the relevant graphical information and tabulation of input and results. Reference should also be made to any third-party reports such as core analysis and petrographic studies.

8
Beyond Log Analysis

8.1 Pressure measurements, gradients and contacts

There are three basic descriptions of pressure in a borehole: hydrostatic, overburden and formation pressure (Figure 8.1):

- *Hydrostatic* pressures are those due to a connected fluid column from the surface to a given depth on the subsurface; the hydrostatic gradient is a function of depth and fluid density and is also known as 'normal pressure'.
- *Overburden* pressure is the total pressure exerted by the weight of the overlying rock and the pressure of the formation fluids; the overburden gradient is a function of the bulk density and height of the rock column. This is also known as the principle stress direction in geomechanical terms.
- *Formation* pressure is the pressure of the fluids contained in the pore spaces of the sediments and can be 'normal' or 'subnormal'/'abnormal', also termed underpressured or overpressured, respectively; both may be hazardous during drilling.

For all three calculations, estimates of the fluid density and bulk density are required and wireline measurements provide both of these variables either directly or indirectly. Other data can be gathered during drilling to calibrate these formation parameters, such as cuttings, gas measurements and LWD logs.

Formation pressure measurements are readily made using wireline tools with the generic name wireline formation tester (WFT) (Figure 8.2); specific tools are the repeat formation tester (RFT), the formation tester (FMT) and the modular formation dynamics tester (MDT), which is one of the latest generations of multi-probe tools that can measure cross-flow between layers in the reservoir. These tools basically insert a metal probe into the borehole wall to measure the static pressure, induce a pressure drawdown and measure the response of the

Petrophysics: A Practical Guide, First Edition. Steve Cannon.
© 2016 John Wiley & Sons, Ltd. Published 2016 by John Wiley & Sons, Ltd.

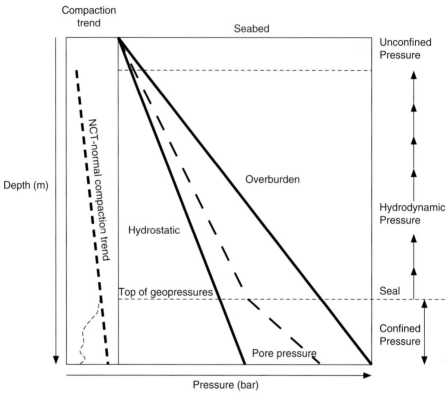

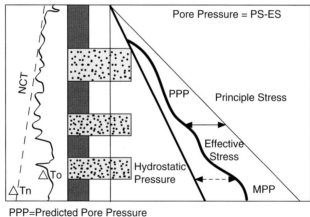

PPP=Predicted Pore Pressure
MPP=Maximum Pore Pressure

Figure 8.1   Different formation pressure gradients with increasing depth. Source: adapted from Shaker (2007). Reproduced by permission of the Canadian Society of Exploration Geophysicists.

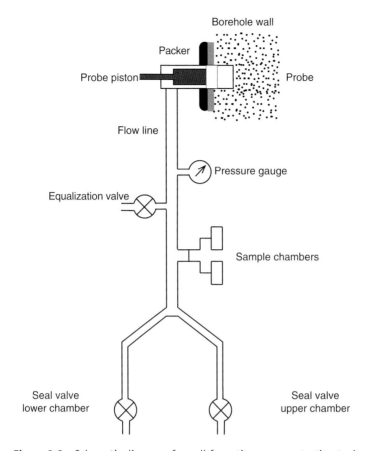

Figure 8.2 Schematic diagram of a well formation pressure-testing tool.

pressure build-up; they are also designed to collect single or multiple formation fluid samples. From the pressure build-up and also the ease of sampling, a qualitative estimate of permeability can be made. The measurements are subject to many ambiguities, especially in low-permeability zones or where the borehole is badly washed out. Using the results of WFT measurements requires care and attention to the basic data acquisition parameters: depth, pressure datum, environmental conditions and the accuracy or precision of the gauges used.

It is always important to consider the results within their stratigraphic context, especially when comparing data from different wells; are observed pressure differences explainable by local barriers, for instance, or is there an indication of reservoir compartmentalization due to faulting? All pressures must be recorded in consistent units, either bars per metre (bar/m$_a$ or bar/m$_g$) or pounds per square inch (psi$_a$ or psi$_g$: where subscript 'a' stands for atmospheric and 'g' for gauge). This simple datum variation of 14.7 psi has been the cause of numerous mistakes in the integration of data. Another simple quality check is to ensure that the mud hydrostatic pressures taken before and after the formation pressure test are the same; if not, then the tool is reading inaccurately. After performing a build-up test or sampling, it is advisable to allow the pressure to restabilize to the initial value before withdrawing the probe. If the formation is permeable, this stabilization time may be fairly short and acceptable to the driller, but extended pressure tests run the risk of the tools sticking irretrievably. The accuracy of the pressure measurement can be a function of the type of gauge used; two gauge types are commonly used: strain gauges and quartz gauges. Strain gauges have a resolution of about 1 psi, but the accuracy of a 10,000 psi rated gauge at full deflection is ±15–20 psi. High-precision quartz gauges are more accurate at about ±1 psi on a 10,000 psi measurement, but this is dependent on knowing the bottom-hole temperature accurately. A further disadvantage is the longer stabilization time required by the more sensitive tool.

There is a special phenomenon known as *supercharging* that results in pressure readings that are substantially higher than expected. This is seen most commonly in low-permeability formations that have been drilled with overbalanced mud. Supercharging is recognized as pressure readings above the expected and tending towards the mud hydrostatic pressure; essentially, the mud is continuing to invade the borehole wall and 'pressure-up' the formation rather than forming an impermeable mudcake; when reviewing suspect data, look out for terms such as 'tight', 'low permeability', 'slow build-up' or 'test aborted' in the Comments section.

Having established a reliable and consistent set of pressure data, it can be used to investigate fluid gradients in single or multiple wells through simple graphing methods (Figure 8.3); always use the true vertical depth measurement to compare wells. Having plotted the pressure data correctly against depth, one or more trends may be identified that will relate to gas, oil or water gradients. Slopes with a gradient of 0.0233–0.032 bar/m (<0.31 psi/ft) are likely to represent gas or gas condensate, 0.069–0.087 bar/m (0.32–0.36 psi/ft) light oil and 0.37–0.41 psi/ft heavier oil, and 0.433 psi/ft is the gradient for fresh water.

These gradients can be related back to the specific gravity of the liquid through the equation

$$\text{Gradient}: \text{psi/ft} = \text{specific gravity}/2.31 = 0.09806 \text{ b/m}$$

where specific gravity is in g/cm^3. The relationship between specific gravity and oil density in API degrees is given by

$$\text{Oil density at surface conditions}: \text{API} = (141.5/\text{specific gravity}) - 131.5$$

As the gas-to-oil ratio of the liquid increases, the apparent density of the oil at reservoir conditions will be reduced, leading to a lowering of the RFT gradient.

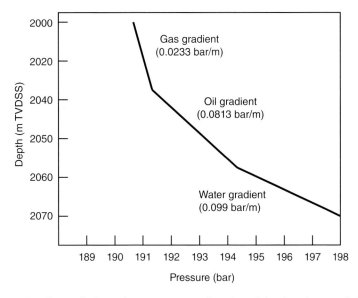

Figure 8.3 Change in formation pressure as a function of depth and reservoir fluid.

Table 8.1 Typical pressure gradients and fluid densities.

Fluid	Gradient (bar/m)	Gradient (psi/ft)	Density (g/cm³)
Dry gas	0.022	0.100	0.230
Wet gas	0.032	0.140	0.320
Oil limit	0.069	0.300	0.689
Light oil 60 API	0.087	0.387	0.780
Heavy oil 20 API	0.091	0.404	0.934
Fresh water	0.098	0.433	1.00
Sea water	0.101	0.444	1.025

Gradients in the aquifer can be used to estimate formation water salinity if R_w and BHT are known, using the standard conversion chart. For example: a water gradient of 0.45 psi/ft is equivalent to a specific gravity of 1.04 g/cm³, which is translated as a salinity of 60,000 ppm NaCl at 200 °F (93 °C) or a resistivity of 0.046 Ω m.

Typical pressure gradients and fluid densities are given in Table 8.1.

8.2 Saturation-height functions

To represent the fluid distribution of a hydrocarbon reservoir in a 3D model correctly, it is necessary to consider honouring the physics of the system. This is done for a given class of rock at a height above a datum where capillary pressure is zero and water saturation is 100%; this is the saturation–height relationship and the datum is called the free water level. Often the results are distributed using the stratigraphic zonation of the reservoir model. There are two primary sources of data for saturation–height modelling: core-derived capillary pressure measurements and saturation estimates from log data.

As previously described, capillary pressure (P_c) is expressed in terms of the interfacial tension between the wetting and non-wetting fluid phases, σ, and the contact angle between the wetting phase and the rock surface, θ, as follows:

$$P_c = \frac{2\sigma \cos\theta}{r}$$

where r is the effective pore radius.

In the laboratory, the fluids involved are either water and mercury or simulated oil and brine, depending on the experimental method, and therefore need to be converted to reservoir conditions using the following equation and conversion values presented in Figure 8.4:

$$P_{cRes} = \frac{P_{cLab}(\sigma \cos\theta)_{Res}}{(\sigma \cos\theta)_{Lab}}$$

The height-related capillary pressure data are related to a saturation–height (H) relationship using the following equation:

$$H = \frac{P_{cRes}}{g(\rho_1 - \rho_2)}$$

where g is the gravitational constant and ρ_1 and ρ_2 are the densities of water and hydrocarbon, respectively (Figure 8.4). Capillary pressure is a function of pore throat size, rather than pore volume, and is therefore subject to the effects of confining pressure at reservoir conditions. Care should always be taken when building a database of capillary pressure data to ensure consistency in experimental methods and conditions.

Before being able to develop a log-derived saturation–height relationship, it is necessary to convert measured depth values to true vertical depth sub-sea (TVDSS). This is best done using a directional survey and the appropriate algorithm in the log analysis package. The log saturation data should only include good-quality data from the cleanest and thickest sands to eliminate uncertainty over clay-bound water and shoulder effects of the input resistivity logs.

Worthington (2002) identified three categories of saturation–height relationship; single- and multi-predictor algorithms and normalized functions.

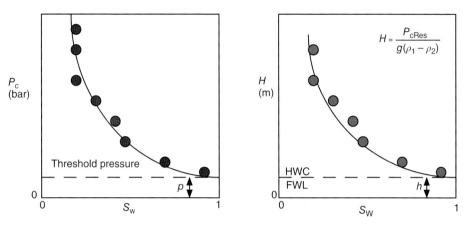

Laboratory conditions		
System	Contact angle θ (degrees)	Interfacial tension σ (dynes/cm)
Air-water	0	72
Oil-water	30	48
Mercury-air	140	480

Generic reservoir conditions		
System	Contact angle θ (degrees)	Interfacial tension σ (dynes/cm)
Gas-water	0	50*
Oil-water	30	30

Figure 8.4 Conversion of laboratory capillary pressure data to reservoir conditions. Source: after Worthington (2002). Reprinted by permission of the AAPG, whose permission is required for further use.

Ideally, each rock type should have a unique saturation–height relationship based on either geological or petrophysical properties. Cannon (1994) coined the term 'petrofacies' to establish a link between geological and petrophysical attributes of a unique, log-derived facies predictor, characterized by definitive mean values of porosity, permeability and water saturation.

8.2.1 Single-predictor algorithms

This category is the simplest, using only height as a predictor of water saturation (Skelt and Harrison, 1995):

$$S_w = aH^b \text{ or } \log S_w = c \log H + d$$

where a, b and c are regression constants. These simple equations are often used to describe saturation in specific porosity bands or petrofacies; however, they have limitations when applied in 3D models unless conditioned by the geological descriptor also.

8.2.2 Multi-predictor algorithms

These are more complicated algorithms that incorporate porosity and/or permeability in the relationship. Cuddy et al. (1993) proposed a solution that relates height to bulk volume of water (BVW), the product of porosity and water saturation:

$$BVW = aH^{b}$$

or alternatively

$$S_{w} = \frac{aH^{b}}{\phi}$$

where ϕ denotes porosity and a and b are regression constants. If the input variables show a log-normal distribution, such as permeability, then these equations can be rewritten thus:

$$\log BVW = k + a \log H - \log \phi$$

and

$$\log S_{w} = k + a \log H - \log \phi + c \log K$$

where c is a regression constant.

8.2.3 Normalized functions

An example of the third type of relationship is the Leverett-J function (Leverett, 1941) that relates porosity and permeability to saturation through the following equation:

$$J(S_{w}) = \frac{P_{c}}{\sigma \cos \theta} \sqrt{\frac{k}{\phi}}$$

where P_{c} is the pressure differential between the FWL and the measured point $[P_{c} = gH(\rho_{1} - \rho_{2})]$ and the $\sigma \cos \theta$ term represents the surface tension and contact angle from laboratory experiments. When using log-derived saturation

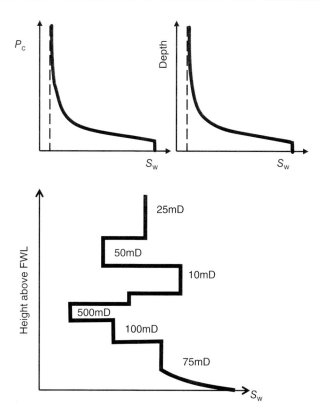

Figure 8.5 Relationship between capillary pressure, height and permeability, demonstrating the impact of rock type on water saturation.

data as input, it is not necessary to include the contact angle and viscosity terms as the data are already at reservoir conditions, so only height and fluid density are required as additional input.

When plotting saturation against height for different rock types, it becomes apparent that permeability has a marked influence on saturation regardless of the height above the free water level (Figure 8.5).

8.3 Electrofacies and facies analysis

Electrofacies discrimination allows us to extend interpretation of reservoir facies into uncored intervals while trying to honour the link between sedimentology and petrophysics. However, because the approach is based on individual log values at each measurement increment, the changes are often too rapid to be used sensibly in subsequent geological modelling. It is probably of greater value to interpret lithology alone and to allow the geologist to group the results as facies associations representing specific deposition packages.

Most log analysis software products have built-in solutions to determine lithology from different logs either through solving multiple simultaneous equations or by some stochastic solution; these often determine porosity at the same time (Figure 8.6). Some tools have sophisticated statistical methods, including fuzzy logic, cluster analysis, principal component analysis and neural networks, to determine different electrofacies. All of these methods depend on robust input training sets based on a core description if they are to be used successfully; without calibration to core or cuttings, the results cannot be validated and should be treated with scepticism. Even where input data are core constrained, the success of facies recognition is only about 80% correct, and where there is no core about 60% correct, if the training set is well constrained.

In either simple or complex mineralogical/porosity associations, the log responses for any zone may be related to the sum of the proportions of the components, each multiplied by the appropriate response coefficients in a series of simultaneous equations (Doveton, 1994). The equation for each log takes the form

$$c_1 v_1 + c_2 v_2 + \cdots + c_n v_n = l$$

where n = number of logs, v_i = proportion of the ith component, c_i = log response of the ith component and l = log response of the zone. For example, in a limestone–dolomite–anhydrite–porosity system with density, sonic and CNL logs, the number of components (m) is 4, the number of logs (n) is 3 and the n log equation might be as follows:

$$2.71\, v\_ls + 2.87\, v\_dol + 2.98\, v\_an + 1.00\, PHI = 1\_den$$
$$47.5\, v\_ls + 43.5\, v\_dol + 50.0\, v\_an + 189.00\, PHI = 1\_son$$
$$0.00\, v\_ls + 7.50\, v\_dol - 0.20\, v\_an + 100.00\, PHI = 1\_cnl$$

where PHI = v_por. Because of material balance, the proportions of the components sum to one:

$$v_1 + v_2 + v_3 + \ldots + v_n = 1.00$$

In the example:

$$v\_ls + v\_dol + v\_an + PHI = 1.00$$

In this example there are n = 4 equations (n = 3 for the logs plus the unity equation) and m = 4 unknowns (the proportions of each component). Rewriting these equations in matrix algebraic terms:

$$C \cdot V = L$$

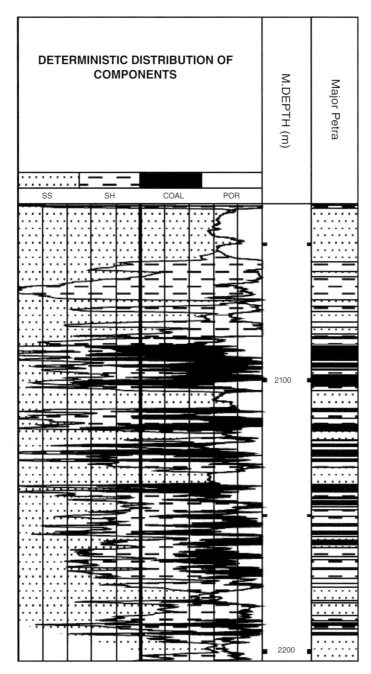

Figure 8.6 A deterministic distribution of lithological components and interpretation of results using the Petra algorithm in TerraStation.

In the example:

C			V	L
2.71	2.87	2.98	Vls	1den
47.50	43.50	50.00	Vdol	1son
00.00	7.50	−0.20	Van	1cnl
1.00	1.00	1.00	PHI	1.0

The matrix formulation is a linear model and generally provides a satisfactory first approximation for compositional solutions, as confirmed by core analysis and laboratory studies. However, the borehole environment, the tool design characteristics and the physics of the measurement variable introduce sources of non-linearity. Local-fit linear models can easily accommodate obvious non-linear functions, such as those relating the pore volume to the neutron response.

The matrix algebra concisely relates the information available and the degree of complexity of the component association. Each log provides a single equation and the collective set is supplemented by the material balance equation (unity equation). When the number of logs is n, the number of equations is $n + 1$, which can be used to resolve $n + 1$ components uniquely; in this situation, the system is 'uniquely determined'. If the number of logs is inadequate, the system is 'underdetermined' and the solutions are downgraded to estimates stipulated by external constraints or by prior geological knowledge. If the number of logs exceeds the minimum requirement, the system is 'overdetermined' and a solution must be chosen that is most consistent with all the available data. The algorithms used to resolve these three possible states of determinacy have strong structural similarities and are reviewed in Doveton and Cable (1979).

8.4   Rock typing

Rock typing is a process that should combine geological and petrophysical characterization at the core scale that can be applied at the wireline log and even seismic scale. A geological facies or facies association will often comprise more than one rock type, suggesting that rock types are linked to depositional processes at the lithofacies scale. However, it is difficult to characterize, let alone model, every lithofacies described by the sedimentologist, so a pragmatic approach to exercise must be imposed; 5–10 rock types should be sufficient to characterize most reservoir types and fewer is often better. The term petrofacies has been coined for these geological constrained rock types (Cannon, 1994).

A simple workflow for rock typing has been established that can work in both clastic and carbonate reservoirs:

1. Describe core and use petrography to classify core plugs by dominant lithology and texture; look for zone patterns.
2. Group core analysis points by dominant lithology and plot porosity/permeability; identify discrete porosity classes.
3. Perform regression analysis on discrete data 'clouds'.
4. Integrate P_c data where available to determine pore-size distributions for discrete lithologies.
5. QC results by comparing actual versus predicted permeability.
6. Correct log derived porosity (ϕ_{eff}) with core porosity (QQ plot).

For example, in a fluvial reservoir there may be the following facies or facies associations identified:

Facies association	Facies		
Channel (30%)	Channel lag (5%)	Active channel fill (10%)	Channel abandonment (10%)
Overbank (20%)	Minor channel	Crevasse splay	
Flood plain (50%)	Hetrolithics	Soils, coals, etc.	

However the floodplain deposits may be considered as one non-contributory rock type in terms of storage capacity and flow potential; in reservoir modelling terms this is often 'background'; these deposits tend to have the highest gamma ray response. The channel deposits can be characterized by three capillary pressure profiles representing different pore geometries, but can also be recognized in core and on log as 2–3 m thick sand bodies with a blocky profile at the base becoming bell-shaped towards the top as the grain size becomes finer. Overbank deposits are associated with both channel and floodplain facies but will be represented by thinner channel bodies and discrete sandy events. It may be possible to group petrophysical rock types from different facies associations where their capillary pressure profiles are similar.

An all-encompassing concept to capture property variability in a facies model is that of the (hydraulic) flow zone: unfortunately, this concept can mean different things to different people. A hydraulic flow zone is related to the geological facies distribution but may differ in terms of boundaries and is seldom vertically contiguous. It is defined by geological characteristics such as texture and mineralogy and petrophysical properties related to them – porosity, permeability and capillary pressure. A hydraulic zone is defined as 'the representative elementary volume (REV) of total reservoir rock within which geological and petrophysical properties that may affect fluid flow are internally consistent and predictably different from other rock volumes' (Amaefule et al., 1993). But what does this mean to different disciplines?

- To a geologist – it is a definable 3D facies object such as a fluvial channel or a carbonate shoal.
- To a petrophysicist – it is a 2D correlatable zone with similar petrophysical properties.
- To a reservoir engineer – it is a 3D reservoir layer that has a consistent dynamic response in the reservoir simulator.
- To a reservoir modeller – it is all these things!

Amaefule et al. (1993) presented a robust and effective way of integrating core and log data to characterize better the fluid flow in a reservoir. Their approach is used to identify hydraulic (flow) units and to predict permeability in uncored intervals or wells, while retaining a link to depositional facies. The classic discrimination of geological rock types has been based on observation and on empirical relationships established between porosity and the logarithm of permeability, but often permeability can vary by 1–3 orders of magnitude over the same porosity range. There is no physical relationship between porosity and permeability; rather, permeability is dependent on grain size and sorting and therefore pore throat distribution. Their method uses a modified Kozeny–Carman equation and the concept of mean hydraulic radius. The equation indicates that for any hydraulic unit a log–log plot of reservoir quality (RQI) against normalized porosity (PhiZ or ϕ_z) should produce a straight line with unit slope, whose intercept where porosity is equal to unity defines a unique flow zone indicator (FZI) for each hydraulic unit. The defining terms are all based on overburden-corrected core-derived porosity and permeability data.

Without going into all the theory, the practical application is relatively straightforward once a simple facies model has been established.

1. Cross plot core porosity against core permeability and define one or more correlation lines related to the facies identified. Try to apply these relationship(s) through the uncored interval in a well to test their robustness.
2. Create the log–log plot of RQI versus PhiZ and compare with the previously defined facies; is there correspondence? Can the relationship be used predictively?

$$\text{RQI} = 0.0314\sqrt{\frac{k}{\phi_e}}$$

$$\text{PhiZ}\left(\phi_z\right) = \frac{\phi_e}{1-\phi_e}$$

$$\text{FZI} = \frac{RQI}{\phi_z}$$

On the log plot of RQI against PhiZ, all samples with similar FZI values will lie on a straight line with unit slope, having similar pore throat distributions and thus belong to the same hydraulic unit.

3. Permeability is calculated using an appropriate hydraulic unit relationship with its mean FZI value and porosity. This can be an iterative process, as the results should converge with the core data.

$$k = 1041\text{FZI}^2 \left[\frac{\phi_e^{\,2}}{\left(1-\phi_e\right)^2} \right]$$

4. Once the porosity–permeability relationship has been robustly established for each facies or rock type on the core data, it can be applied to the log-derived porosity data in all wells, again by facies or rock type.

5. Where capillary pressure data are available, these hydraulic units should also describe different rock types based on the wetting phase saturation values. This means that each rock type can also be characterized in terms of a saturation–height relationship.

For a fully worked-out example, see Corbett and Potter (2004).

8.5 Integration with seismic

The borehole-compensated sonic log (BHC) of the 1980s has largely been replaced by a range of different sonic tools designed to capture full wave-form data; the sonic log is seldom used for porosity determination now, but is the main tool of the geophysicist developing 'petro-acoustic' models of a reservoir. These days, the petrophysicist is expected to provide prepared log data to geophysicists for the depth conversion of interpreted horizons and also for fluid substitution studies and seismic attribute modelling. Of these, the first and last have the greatest impact on the reservoir modelling process and volumetric estimation. A poorly constrained depth conversion can have a major impact on the GRV model as it moves in space with respect to the base level, usually the hydrocarbon fluid contact. The GRV on a field I once worked on was found to 30% smaller after a reinterpretation of the seismic and model rebuild. The error was found in the depth conversion of the top reservoir horizon: only platform wells had been used to constrain the depth conversion and, as a result, the time–depth relationship in the overburden did not vary away from the platform. The flanks of the depth structure were therefore incorrectly represented in the depth conversion and the structure had collapsed inwards.

8.5.1   Depth conversion

Seismic data are in time: to make them available for use in well planning or volumetric modelling, they must be converted to depth. The depth conversion workflow is based around the creation of a synthetic seismogram used to tie wireline formation tops with an interpreted seismic horizon: this is done at several levels in the overburden and above the reservoir zones. A synthetic seismogram has two components: the generation of an acoustic impedance (AI) log and conversion of this log to time such that it can be compared with the seismic profile at the well. Another product of this process is the creation of a reflectivity log, which is what is actually compared with the seismic section.

An AI log is the product of the sonic and density data from the well: the sonic data are converted from transit time (μs/m) to velocity (m/s) and the density log is converted from g/cm^3 to kg/m^3, so that AI is reported in kg/m^2/s. Both the sonic and density logs must first be corrected for washouts, fluid invasion and depth discrepancies and datumed with respect to the seismic, usually sub-sea depth for offshore data. The AI log must then be converted to two-way time to be in the same 'space' as the seismic volume. This is usually done by mathematical integration of the sonic log and calibrated with any well seismic data such as a check-shot or vertical seismic profile (VSP) acquired in the well. The next step is to create a reflectivity curve from the impedance data, done by differentiating the log with respect to time. The reflectivity (R) is defined as

$$R = \frac{AI_1 - AI_2}{AI_2 + AI_1}$$

where AI_1 and AI_2 are adjacent data values in the well (Figure 8.7).

With both AI and R defined in time (t), to make them comparable to the seismic it is necessary to convolve the logs with a seismic wavelet that is representative of the phase and frequency content of the seismic volume: commonly this will be either a zero or minimum phase wavelet. Selection of the correct wavelet is often a matter of trial and error, looking for the best match between a strong seismic event and a strong well marker that represent the same horizon in the subsurface. Increases in AI produce positive or 'hard' events that indicate a change in lithology, say from shale (soft) to sand (hard), whereas decreases in AI create 'soft' events: SEG convention would normally have hard events (black) kick to the left and soft (white) to the right, although

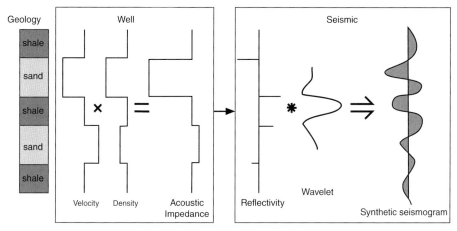

Reflection Coefficient = $(AI_2 - AI_1)/(AI_2 + AI_1)$

Figure 8.7 Process diagram to establish a well-to-seismic correlation. Source: adapted from Schroeder (2006). Reprinted by permission of the AAPG, whose permission is required for further use.

this is often a matter of choice. In the North Sea, the opposite convention is normal, but not in all companies. It may be necessary to shift the logs to accommodate seismic artefacts in the overburden, such as a gas cloud, surface channel or igneous intrusion, that impact on the overall well tie, but within reason this is a normal part of the process. Whether or not this works is dependent on the size of the AI contrasts and the quality of the log analysis and seismic interpretation.

The actual process of depth conversion using the well tie information is the work of the geophysicist, but the process can be made easier if the velocity log data provided have been thoroughly reviewed and presented in an easily recognizable format. Perhaps by working with the geophysicist it is possible to provide the correct interval velocity on a well-by-well basis and to identify anomalies together. One word of warning: the seismic pick and the geological pick may not be exactly the same event and can suffer from the poorer resolution of the seismic data; a compromise is usually required to achieve the best fit of the two data types, so include the geologist in the discussion!

8.5.2 Fluid substitution

Fluid substitution is an important part of rock physics analysis that provides a way of identifying and quantifying what fluids occupy the pore space of a reservoir. The objective of fluid substitution is to model the seismic velocity and density of a reservoir under *in situ* conditions (pressure and temperature) and with knowledge of the porosity, mineralogy and formation water salinity. Oil or gas in a reservoir will normally reduce the response of the compressional sonic velocity compared with brine-filled formations; the effect is most apparent in the gas-filled case, such that even low concentrations of gas will give a strong impedance response. The seismic velocity of an isotropic material can be estimated from the rock moduli and density using the compressional (V_p) and shear (V_s) wave sonic data acquired from logs:

$$V_p = \sqrt{\frac{K + \dfrac{4\mu}{3}}{\rho}}$$

$$V_s = \sqrt{\frac{\mu}{\rho}}$$

The most commonly used approach to fluid substitution is to use the Gassmann equations developed in the 1950s (Gassmann, 1951). Modelling the changes from one fluid type to another in the reservoir implies that the effects of the original fluid must be removed so that the bulk and shear moduli and bulk density of the unit under investigation can be calculated. The bulk modulus (K) is a measure of the resistance to volume change under an applied stress and the shear modulus (μ) resistance to a change in shape. When the new fluid is substituted, the bulk modulus and bulk density can be recalculated and hence the compressional and shear velocity for the new conditions. At its simplest, the Gassmann equation states that the bulk modulus (K_{sat}) of a rock saturated with fluid of bulk modulus (K_{fl}) is

$$\frac{K_{sat}}{K_{ma} - K_{sat}} = \frac{K_d}{K_{ma} - K_d} + \frac{K_{fl}}{\phi(K_{ma} - K_{fl})}$$

where K_{ma} is the bulk modulus of the matrix, K_d is the bulk modulus of the dry rock frame and ϕ is the porosity.

It is a function of the Gassmann approach that the shear modulus remains unchanged during the process provided that the reservoir is not fractured or vuggy, such that

$$\mu_{sat} = \mu_d$$

The challenge is deriving some of these terms for the different moduli; calculating the saturated moduli assumes knowledge of the dry rock moduli. Although this can be estimated in the laboratory or derived from empirical equations, if all we want to know is whether a reservoir is filled with oil, gas or water prior to drilling, we just need to know the response in a single-fluid case to calculate the response in the others. Obviously we must have good-quality log data and some understanding of the geology of the reservoir to attach any confidence to the results. In general, the lithology effects are considered to be greater than the fluid effects in the AI; this is because the lithology tends to be more variable over an interpreted section, especially with marked porosity changes. The shear velocity can also be used to create an elastic impedance (EI) trace from logs in the same way as the compressional sonic; this curve or volume is largely independent of fluid.

Applying both EI and AI data can lead to a wealth of information on the reservoir. These relationships, although originally developed to identify reservoir fluids, can also be used for amplitude-versus-offset (AVO) analysis, a tool that has become extremely valuable to exploration in marine Tertiary plays, where the seismic response is most suitable for identifying direct hydrocarbon indictors (DHIs). For a reasonably readable description and explanation of the process and physics behind seismic property analysis, see Bacon et al. (2003).

8.6 Production logging

Once a borehole has been isolated or cased off, acquiring further formation evaluation information becomes more of a technical and economic problem, especially if a well is in production. Typically cased-hole tools either measure reservoir performance or check the integrity of the cement or casing. Tools are based on normal open-hole technology, often with enhanced capability to 'see' through the barrier formed by casing. Neutron porosity, photo-density and gamma ray measurements can be combined to monitor changes in saturation levels and also identify lithology, porosity and zones of by-passed hydrocarbons after a period of production. Acoustic scanning measurements are used to perform high-resolution cement evaluation, casing corrosion determination and *in extremis* borehole image information. Production logging tools (PLTs) are generally simple mechanical spinner mechanisms that measure the rate of flow over a perforated interval to determine what parts of a well are contributing to production. In combination with a robust reservoir description, they can be used to identify key producing rock types.

8.6.1 Pulsed neutron logging

Pulsed neutron logs have not been adopted wholesale for open-hole formation evaluation, mainly because of the vast amount of legacy data from the CNL-type tools. However, the technology has found favour in the world of reservoir monitoring during hydrocarbon production. The typical use of a PNL log is to monitor reservoir saturation through time, responding to the chlorine in the formation brine and thus the change in water salinity. Additionally, these tools can determine the carbon/oxygen ratio and capture spectra and porosity. The tools generally combine a high-yield neutron generator with a high-efficiency dual detector to provide a compensated, through-casing log.

Reservoir saturation information is derived from carbon/oxygen (C/O) ratios or inferred from the rate of capture of fast thermal neutrons – sigma capture. The carbon/oxygen ratio is measured in two ways: either the C/O yield is obtained from a full spectral analysis or by placing a broad window over the carbon and oxygen spectral peaks of the inelastic spectrum of readings. Because of the type of statistical processing used to get a result, the C/O yield is more accurate but less precise than the spectral window approach; both methods are used to obtain the best estimate of hydrocarbon volumes in the pores. By using C/O ratios from the near and far detectors, water saturation and borehole hold-up may be obtained.

The sigma capture method is a measure of how fast thermal neutrons are captured, a process dominated by chlorine, present as dissolved salts in the formation water and drilling fluid. To overcome the influence of drilling fluid on the measurement, the neutron generator is pulsed in a dual burst pattern: a short burst followed by a long burst. The near detector 'sees' the borehole fluid, whereas the far detector 'sees' the formation sigma, especially from the long neutron burst. Other environmental corrections are required to obtain a final processed saturation profile; these corrections for casing, hole size, salinity, porosity and lithology are derived from analogue data.

The real value in these measurements becomes apparent when they are repeated after a period of hydrocarbon production, giving a 'time-lapse' set of information, which although not necessarily accurate will at least show how water saturation has changed and any movement in the oil–water contact.

8.7   Geo-steering

Geo-steering is in essence drilling a well that is located within the hydrocarbon-bearing, porous section of a reservoir to maximize the potential completion intervals of a well. The development of steerable bottom-hole assemblies, mud motors and measurement-while-drilling tools, led to the advent of extended reach drilling (ERD) and designer wells. To keep the drill bit within the target tolerances of the geologist required advanced techniques such as wellsite biostratigraphy and chemical stratigraphy. LWD tools were also developed into azimuthal versions, especially for resistivity measurements that could 'see' when the drill bit was nearing a boundary between the reservoir sand and shale or when approaching the hydrocarbon–water contact. This would allow the drill to steer the drill bit to keep it within the desired reservoir unit with sufficient standoff from the contact. Azimuthal density tools are also used for geo-steering to ensure that the most porous intervals are drilled. A combination of wellsite geological analyses and LWD information leads to a successful geo-steered well, but both suites of information are transferred through the mud column as cuttings or data pulse and are therefore 'lagged' some time after the drill bit has penetrated a section. There is therefore a dependence on rapid data transfer and interpretation of results.

8.8   Petrophysics of unconventional reservoirs

Unconventional reservoirs come in many forms, for example, tight gas sands, fractured basement granites and increasingly shale gas and shale oil reservoirs. The common ingredient is that it is difficult to estimate the value of the potential resource using industry standard methods. Taking shale reservoirs as an example, the main questions that need to be answered are the proportion of organic material that might generate hydrocarbons, the hydrocarbon pore volume, the areal extent and the possible production mechanisms. Petrophysics can try to address some of these issues such as total organic content (TOC), maturity, gas or oil content, water saturation and porosity and the geomechanical properties of the rock, specifically fracturing. As ever, this is often difficult to do because of the type, age and quantity of data – either too much or too little! It is difficult to estimate TOC from log data without laboratory calibration data, and porosity estimation is complicated by the presence of organic material. Organic-rich rocks are considered to be comprised of three components: rock matrix, solid organic matter and fluids filling the pore space (Figure 8.8).

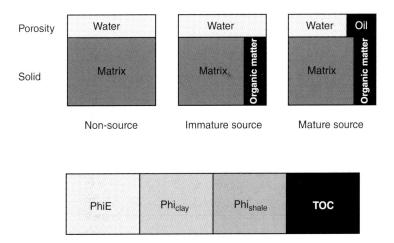

Figure 8.8 Components of source rocks and non-source rocks and the four types of porosity found in shale rocks.

8.8.1 Total organic content

A practical model for estimating organic richness from porosity and resistivity logs was developed by Passey et al. (1990) and has been the mainstay of understanding potential shale gas reservoirs ever since. The method is called the $\Delta \log R$ technique and is essentially an overlay method in which a correctly scaled porosity log, usually sonic porosity, is plotted with a deep-reading resistivity log: the sonic should be scaled at 50 µs/ft for each decade of the resistivity log in Ω m. In either hydrocarbon-bearing reservoir rocks or organic-rich non-reservoir rocks, a separation between the curves occurs. The curve separation is a function of low-velocity kerogen in the rock, while the resistivity is responding to the formation fluid (Figure 8.9). Combined with a gamma ray, preferably the spectral tool, the presence of uranium-rich marine shales can be identified. This approach may also be applied to a cross-plot display of the data to compare data from multiple wells quickly.

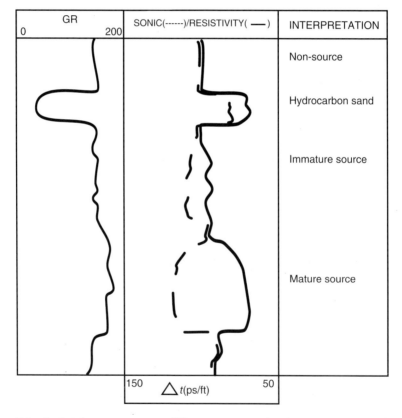

Figure 8.9 Typical log responses to different source rocks. Source: Passey et al., 1991. Reprinted by permission of the AAPG whose permission is required for further use.

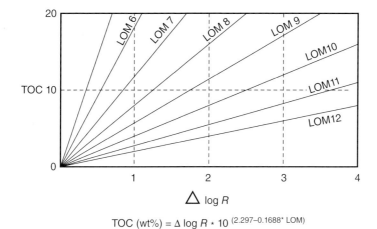

$$TOC\ (wt\%) = \Delta\log R * 10^{(2.297-0.1688^* LOM)}$$

Figure 8.10 Graph of TOC against $\Delta\log R$ to estimate levels of maturity (LOMs) of shale source rocks. Source: Passey et al. (1990). Reprinted by permission of the AAPG, whose permission is required for further use.

The $\Delta\log R$ relationship is expressed as

$$\Delta\log R = \log\left(\frac{R}{R_{\text{baseline}}}\right) + 0.02\left(\Delta t_{\text{baseline}}\right)$$

where the baseline values for resistivity and shale are picked as non-organic shales. At any given $\Delta\log R$, TOC decreases as the level of organic maturity (LOM) increases (Figure 8.10). The equation can thus be expressed in terms of TOC and LOM:

$$TOC\left(wt\%\right) = \Delta\log R \times 10^{(2.297-0.1688\text{LOM})}$$

The TOC and LOM are laboratory-derived measurements, usually made using the Rock-Eval analytical equipment. It should also be noted that $\Delta\log R$ does not work at high levels of maturity as there has not been sufficient calibration at these higher values.

TOC can also be estimated from a linear relationship with total gamma ray and the bulk density log (Schmoker, 1981). In organic-rich marine rocks, the gamma ray is responding to high concentrations of uranium-rich organometallic complexes that form in poorly oxygenated, anoxic conditions. The gamma ray logs must be normalized and are often only applicable on a well-by-well basis; the relationship can be shown on a plot of GR (API) against TOC (wt%) to take the form:

$$TOC = mGR + b$$

where m is the slope of the line and b is the GR intercept at 0% TOC.

8.8.2   Porosity estimation

Kerogen, the dominant organic material in source rocks and shale reservoirs, has a density of ~1.15 g/cm^3, which is only slightly higher than that of fresh water. As they both also contain hydrogen, distinguishing them from logs becomes a non-trivial matter. A neutron-density cross-plot approach is a good starting point, but if there are high gas saturations then this becomes less useful.

Another solution requires core porosity and laboratory TOC data to calibrate the results of an iterative approach using variable matrix values; this gives a non-unique solution. Plotting TOC against bulk density gives an apparent matrix value that can then be used to calculate equivalent density porosity. These routines are now built into deterministic and stochastic workflows within most log analysis packages. The advent of geochemical elemental analysis tools by the service companies has led to more advanced solvers that combine the various logs and stochastically solve for lithology, porosity, water saturation and, most importantly, volume of kerogen.

8.8.3   Gas in place

Finally, to enable us to calculate gas in place, we need to understand how the gas is distributed. Adsorbed gas is dependent on the TOC, reservoir pressure, temperature, maturity and kerogen type: adsorbed gas is measured in standard cubic feet per ton of rock. An empirical model using isotherms calibrated to core data from the reservoir leads to a set of Langmuir isothermal equations, comparing TOC with gas content for a given pressure value. The Langmuir isotherm describes the adsorption of an adsorbate, in this gas, on to the surface of an adsorbant, the matrix rock.

Free gas content is calculated using standard analyses to estimate porosity and water saturation in the remaining reservoir.

9
Carbonate Reservoir Evaluation

Carbonate reservoirs, limestone and dolomite in the main, are very much more difficult to analyse petrophysically than clastic rocks because of the more complex pore structures and networks commonly encountered. Carbonate minerals are generally less stable than quartz and are altered more readily during diagenesis, resulting in irregular and unpredictable pore geometry. These in turn affect the relationships established previously, especially for formation resistivity and thus water saturation. Although the normal methods of log analysis are satisfactory for a simple intergranular or intercrystalline carbonate, they are less successful in vuggy, moldic or fractured limestone. It is often necessary to apply other interpretation strategies to unlock the hydrocarbon potential of carbonate reservoirs. A complete review of carbonate reservoir characterization was published by Lucia (1999).

9.1 Rock fabric classification

Carbonates are often classified in terms of their lithology, limestone or dolomite, their bioclastic and chemical components resulting in grain types and size and also pore type. In each case, the dominant petrophysical element is the pore type; however, to integrate geology it is necessary to consider the depositional environment and post-depositional history of the reservoir to understand the controls on porosity and permeability.

Carbonates fall into three pore types: interparticle pores, separate vuggy pores and interconnected vuggy pores and fractures. Each class has a different pore-size distribution and connectivity, resulting in different pathways in the rock fabric through which an electrical current can pass, thus making carbonate rocks highly sensitive in terms of the Archie cementation and saturation

Petrophysics: A Practical Guide, First Edition. Steve Cannon.
© 2016 John Wiley & Sons, Ltd. Published 2016 by John Wiley & Sons, Ltd.

exponents *m* and *n*. As logged formation resistivity changes in a carbonate reservoir, the challenge is to determine whether this change is a function of the water saturation or the porosity of the formation; in other words, is it the *m* or *n* term in the saturation equation that is varying, or both?

Interparticle pores can be described in terms of particle size and sorting and the resultant porosity. Rock fabrics are described as being grain or mud dominated (also termed matrix dominated) in some cases. Particle size can be related to capillary pressure measurements to establish pore-size distributions, the largest pores having the lowest displacement pressure. Lucia (1983, 1999) established that important displacement pressure boundaries could be established at 20 and 100 μm that define three separate permeability fields (Figure 9.1):

- *Class 1 (>100 μm)* – limestone and dolomitized grainstone; large crystalline grain-dominated dolo-packstone and muddy dolostones.
- *Class 2 (20–100 μm)* – grain-dominated packstone; fine–medium crystalline grain-dominated dolo-packstone; medium crystalline muddy dolostones.
- *Class 3 (<20 μm)* – mud-dominated fabrics and fine crystalline dolostones.

These may be incorporated with the rock fabric description to characterize non-vuggy lithologies better. These relationships may be used for limestone and dolomite, or crystalline carbonates.

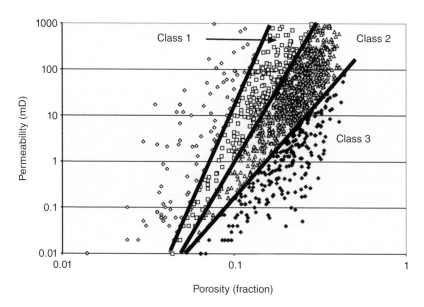

Figure 9.1   Carbonate rock type classification based on Lucia (1999). The example shown is from a non-vuggy dolomitic limestone and important displacement pressure boundaries are be established at 20 and 100 μm pore-throat sizes that define three separate permeability fields.

Vugs are usually the product of dissolution of fossil fragments or depositional grains and may be isolated or connected. Lucia (1983) defined these as separate or touching vuggy pores; separate vugs are only connected through the interparticle pore network while touching vugs form an interconnected pore network independent of interparticle porosity. Separate vugs are usually fabric selective and include intra-fossil pore spaces such as closed bivalve shells, dissolved grains (oomolds) or crystals and intragrain microporosity. Touching vug pore systems are generally non-fabric selective and significantly larger than the original particle size, forming extensive pore networks, including caverns, collapse breccia and fracture systems.

9.2 Petrophysical interpretation

9.2.1 Porosity

In carbonates, the neutron and density tools give a representative estimate of total porosity; the sonic log measures the connected porosity only and hence will give a lower porosity estimate in vuggy rock fabrics. In all other aspects the tools work in the same way and the results are interpreted using the same workflow; however, the need for core calibration is even greater in carbonate reservoirs because of the greater variety of lithologies, rock types and pore systems. The gamma ray response in carbonates is typically <20 API and generally lacking in character, unless there are discrete shale layers. As a result, lithology determination is usually best made with the density log as each of the components has a distinctive bulk density value between 2 and 3 g/cm^3. The neutron log will give a lithology-independent porosity value based on the presence of hydrogen alone; care needs to taken with complex mixed lithologies to obtain the correct proportion of each mineral component, especially if salts are anticipated.

Gypsum ($CaSO_4.2H_2O$) is a hydrated calcium sulfate and the precursor of anhydrite in evaporite sequences. The neutron porosity log will respond to the water of crystallization in the gypsum and give an erroneous, higher porosity; the density of anhydrite is significantly greater than those of many other common minerals and will also give an inaccurate porosity measurement unless the bulk density is corrected for the presence of anhydrite. In non-vuggy carbonates, the standard interpretation methods can be adapted for the different lithologies, e.g.,

$$\text{Bulk density}\left(\rho_{bulk}\right) = \rho_{fluid}\,\phi + \left(2.71V_c + 2.84V_d + 2.98V_a + 2.35V_g + 2.65V_q\right)$$

where V_c, V_d, V_a, V_g and V_q represent the varying volume proportions of limestone, dolomite, anhydrite, gypsum and quartz, respectively, that make up the rock matrix. (See Table 3.3 for a listing of the major minerals and their densities and photoelectric responses.)

9.2.2 Water saturation

In a carbonate with intergranular or intercrystalline porosity, water saturation calculated using the Archie equation is generally reliable; where different porosity types are present, other methods need to be employed to determine hydrocarbon saturation. These include the use of an additional textural parameter (W) and log-derived approaches such as bulk-volume water calculations, production ratio index and moveable hydrocarbon index. The real challenge with carbonates is that their rock fabric and pore type distribution can change rapidly and unpredictably, making a single solution difficult to implement.

A variable m value in a well can be calculated for each interpretation level if both neutron–density and sonic logs are available. The neutron–density combination estimates total porosity in the reservoir and the sonic log estimates connected porosity, in this case vuggy porosity. Nugent et al. (1978) developed the following relationship that works well in carbonates with vuggy and moldic porosity:

$$m \geq \frac{2 \log \phi_s}{\phi_t}$$

where ϕ_s and ϕ_t are the sonic log porosity and the neutron–density (total) porosity log, respectively. A separate vuggy porosity can be estimated from the equation $\phi_{vug} = 2(\phi_t - \phi_s)$ for use in oomolds (Nurmi, 1984), leading to a value for matrix porosity alone:

$$\phi_{matrix} = \phi_t - \phi_{vug}$$

These terms can now be used in the previous equation to calculate the variable m on a level-by-level basis where oomoldic porosity is recognized – a simple rock typing method for carbonates.

Lucia and Conti (1987), working the laboratory and borehole data, further refined this relationship by plotting a series of m values against the 'vug–porosity ratio' (VPR) to give a straight-line relationship:

$$\text{Cementation factor} \, (m) = 2.14 \left(\phi_{sv} / \phi_t \right) + 1.76$$

where ϕ_{sv} = separate vug porosity.

The value of m for non-touching vuggy carbonates can vary between 1.8 and 4 and in the presence of fractures or touching vugs the value can be less than 1.8. If the correct m value is not used in the Archie equation, the water saturation will too high if it is <2 and too low if >2, where 2 is the default value for m in most cases.

Asquith (1985) presented several case studies where an alternative approach to classical log analysis is required in carbonate reservoirs. In the case of the Canyon Reef reservoir, Scurry County, Texas, two separate interpretations had two very different water saturations because the wrong m had been used in the analysis; the default value gave a porosity of 24% and a water saturation of 22%, but in a production test the well flowed water. Using a Picket plot approach combining true resistivity, porosity, water saturation and porosity exponent, it is possible to estimate from the slope of the line a true value for m in the water leg that can then be applied in the analysis. A higher value for m is estimated as 3.7, reflecting a complex vuggy porosity distribution, and when applied to the analysis gives a water saturation of 74% in the reservoir.

Bulk volume water (BVW) calculation is fairly simple, the calculated porosity times the calculated water saturation in a well, and can be used to identify those hydrocarbon-bearing zones at irreducible water saturation (S_{wirr}) and thus likely to flow free from water under production. A reservoir at irreducible water saturation exhibits BVW values that are constant throughout, regardless of the porosity, as all formation water is held through surface tension or capillary pressure exerted by the pore network. Asquith (1985) presented a table comparing BVW with grain size and carbonate porosity type that suggests that at values of BVW <0.04 a reservoir will produce water-free hydrocarbons (Table 9.1). The table also shows that for vuggy reservoirs to produce water free they are required to have very low BVW, because vuggy pores hold little water by surface tension or capillary pressure.

The production ratio index (PRI) method is based on the assumption that vuggy porosity does not contribute greatly to formation resistivity, as the pores are unconnected; matrix porosity controls formation resistivity as measured by the logs (Nugent et al., 1978). Combining resistivity with sonic porosity to estimate the matrix water saturation (using Archie) and then multiplying by the total porosity from the neutron–density cross-plot gives the PRI:

$$PRI = S_{ws} \times \phi_{n-d}$$

A further property of the PRI is that it can be used to estimate the expected initial water-cut at production start-up in carbonate reservoirs characterized with vuggy or moldic porosity: a PRI <0.02 would indicate water-free production whereas a PRI >0.04 would indicate water production only. A combination of these different techniques should be used to understand fully the hydrocarbon distribution in carbonate reservoirs; however, the availability of core data to calibrate both the petrophysical and geological parameters is essential for a successful outcome.

Microporosity in carbonate rocks commonly occurs as a result of micritization of muddy sediments, commonly as in-fill to fossil moulds; diagenesis converts the fine-grained sediment to a crystalline form that holds abundant

Table 9.1   Bulk volume water at irreducible water saturation as a function of grain size and type of carbonate porosity (Asquith, 1985).

Grain size (mm)*		Bulk volume water (BVW)
Coarse	1.0–0.5	0.02–0.025
Medium	0.5–0.25	0.025–0.035
Fine	0.25–0.125	0.035–0.05
Very fine	0.125–0.0625	0.05–0.07
Silt	<0.0625	0.07–0.09
Carbonate porosity type		**Bulk volume water (BVW)**
Vuggy		0.005–0.015
Vuggy and intercrystalline		0.015–0.025
Intercrystalline or intergranular		0.025–0.04
Chalky		0.05

*After Fertl and Vercellino (1978).

bound water. In this case, an electric current flows more easily through the rock than if it were to follow a tortuous pathway around grains or particles. The result is a log response indicative of higher water saturation than in reality, as much of the response is a function of bound water in the micropores. Keith and Pitmann (1983) developed a model of mega- and microporosity to classify different types of carbonate rocks as bimodal or unimodal in terms of the porosity mixture. To do this, they plotted true resistivity (R_t) against the ratio of the flushed zone resistivity and resistivity of the mud filtrate (R_{xo}/R_{mf}). Consistently, carbonates with both porosity types (bimodal) show a lower R_{xo}/R_{mf} ratio than do carbonates with megaporosity; this is a direct result of the higher volumes of bound water associated with the microporosity that cannot be flushed during the drilling process, reducing the reading of the shallow reading resistivity tool.

Guillotte et al. (1979) designed a graphical approach to define a single textural parameter, W, combining the Archie parameters m and n by plotting core-derived porosity and permeability on a log–log plot. Lines of constant W are derived from the following equation relating water saturation and porosity to R_w and R_t:

$$S_w \times \phi = \left(\frac{R_w}{R_t} \right)^{\frac{1}{w}}$$

Values of W will vary from field to field, reservoir to reservoir and rock type to rock type, so careful calibration is required.

From all of the above examples, it is obvious that carbonate reservoirs require special handling to be characterized correctly and that core data are a prerequisite for all approaches. Carbonate petrophysicists have to be imaginative and innovative in the application of the available data to extract the last drop of information and oil from these complex rocks.

10
Petrophysics for Reservoir Modelling

At the start of this book I posed the question, 'What is petrophysics?'. I hope that by now the reader will have grasped what I mean by the subject and also how I see the role of the petrophysicist. Essentially, the petrophysicist is in the centre of the workflow leading to the evaluation and development of oil and gas resources. Nowhere is this more important than in providing the correct input for reservoir modelling, both static and dynamic. The petrophysicist provides sonic data to the geophysicist for depth conversion and for seismic attribute analysis; the petrophysicist supplies the composite logs for well correlation and the relevant porosity, permeability and water saturation input for property modelling, and also helps with the fundamental environment of deposition analysis that leads to the conceptual reservoir model. These are all 'static' properties, but in the dynamic world it is often the petrophysicist rather than the reservoir engineer who defines the various SCAL properties needed to initialize a simulation model, to say nothing about the upscaling of data from the fine to coarse grid. The following sections consider these topics in more detail to show how the petrophysicist can influence and shape a reservoir model and field development plan. For devotees of reservoir modelling I can highly recommend the recent book *Reservoir Model Design* by Ringrose and Bentley (2014): it deals with many of the issues of property modelling and upscaling and also the geological aspects of building a meaningful representation of the subsurface.

Here is another question: 'Why build reservoir models?'. I believe the following are the main reasons, and they are all about describing reservoir uncertainty.

We have incomplete information about the dimensions, architecture and variability of the reservoir at all scales:

- The complex spatial distribution of reservoir building blocks or elements.
- It is difficult to capture the variability in rock properties and the structure of variability with spatial position and direction (anisotropy in properties).

- The unknown relationship between rock property values and the volume for averaging (scale).
- The relative abundance of static point values (porosity, saturation, permeability) over dynamic reservoir data.
- Capacity to integrate data from many disciplines and scales in a convenient and scalable, highly visual representation of the subsurface.

To expand a little on the last point, and to nail my colours to the mast as a geologist first and foremost, many of the data used in reservoir modelling are from regional studies, outcrop analogues, core description and petrography that together provide the context for the rest of the input data. Seismic data, especially time slices or attributes, provide much of the inter-well modelling information and dynamic well-test data, and the proximity to boundaries, but it is the geologists who describe the conceptual model of the reservoir and guide the construction or build it themselves. And the following mantra should be the guiding principle: 'if you can draw it, I can model it!'.

10.1   Multi-scale modelling

A typical hydrocarbon reservoir may have a volume of the order of billions of cubic metres ($\sim 10^9$ m^3), whereas a core or a wireline log taken through that reservoir probably represents only 15–25 m^3 or an order of magnitude difference of $\sim 10^{-13}$. Samples from the core represent a further difference in scale until at the pore scale ($\sim 10^{-12}$ m^3) the order of magnitude difference between a pore and the reservoir is $\sim 10^{-21}$. The dimensions at the well test and seismic scale are obviously not as great as the pore or core data but still amount to significant underrepresentations of the reservoir volume (Figure 10.1).

10.2   Petrophysical issues

When given a suite of interpreted logs to model reservoir properties, it is essential to know what they represent: effective or total porosity, an empirical permeability interpretation based on some software solution. It is important to know whether the data has been core constrained or overburden corrected or whether a Klinkenberg correction has been made to permeability. The petrophysicist should provide all this information, even if the reservoir modeller does not ask for it, or more likely does not know the right questions to ask. Another important question to ask is about the uncertainty inherent in the data: is there sample bias or are there random or systematic errors that need to be corrected? The latter is more common than one might think, especially where more than one operator or service company has drilled and logged the wells.

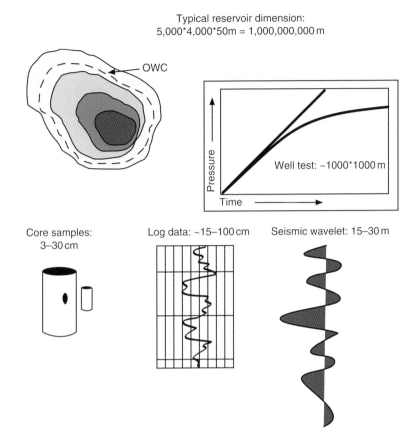

Typical reservoir dimension:
5,000*4,000*50m = 1,000,000,000 m

OWC

Pressure

Well test: ~1000*1000 m

Time

Core samples:
3–30 cm

Log data: ~15–100 cm

Seismic wavelet: 15–30 m

Figure 10.1 Scales of measurement from core data, through log, seismic and well test, to demonstrate the approximately 13 orders of magnitude difference between the different sources.

When modelling reservoir properties, the following basic well information is needed and should be provided, or at least sanctioned, by the petrophysicist whenever possible:

- Mean and standard deviation for each property derived from the well data to be modelled: porosity, water saturation and permeability.
- Cross-correlation between properties: how porosity relates to permeability over a range of values.
- The spatial correlation of properties: how rapidly the property varies with position in the reservoir, i.e. vertical and lateral trends as a reduction in porosity with depth of burial.
- Uncertainty in the conditioning data, either well-derived properties or seismic attributes.

These data should be provided for each reservoir zone to be modelled, both for every well and for the dataset as a whole. When integrated with a facies or rock-type model, these statistical data should also be available.

There is a lot of debate about whether total or effective properties should be modelled and also the use of NTG and property cut-offs in reservoir modelling. My preference is to model overburden-corrected, core-constrained, effective properties because these are representative of the reservoir at depth. In fact, our raw log data are the closest we get to *in situ* reservoir properties. The concept of 'total property' modelling (Ringrose, 2008) is not associated with effective or total properties, but the use of all ranges of data type, without the need for cut-offs or NTG ratios. This approach is most effective when used in conjunction with a facies model: in this case, the non-reservoir component has been taken out of the equation.

10.3   Blocking logs

Log measurements are recorded in 15 cm increments; the minimum cell thickness in a geocellular model is unlikely to be less than 1 m. This means that all petrophysical data require upscaling; transforming the raw data to the geocellular grid process is called 'blocking'; the term 'upscaling' should be reserved for the challenge of moving from geo-grid to simulation grid. Either process attempts to achieve an 'average' representative value for a number of contiguous data points. It is important that the grid design captures major property contrasts seen in the raw data, especially with respect to flow characteristics such as high-permeability streaks or cemented nodules and horizons.

Petrophysical data that require blocking fall into two categories: either discrete or continuous properties. Discrete properties are lithology, facies, rock type and explicit net to gross data; continuous properties are porosity, permeability and water saturation. Discrete properties are usually averaged using the 'most of …' approach, whereby the property that is most often represented (the mode) is chosen. Different averaging methods can be used for continuous properties: porosity and water saturation are usually blocked using an arithmetic average, which may be volume weighted; when upscaling to the simulation grid, an average pore volume property is often used.

Upscaling permeability is the greatest challenge for reservoir property modelling; it is a dynamic property that has a directional component (a vector property). There are many different approaches, ranging from simple averaging to full-scale flow tensor methods. Usually we are struggling to successfully capture permeability variation as it is, so that then also to upscale it smacks of overconfidence! The only reason to upscale permeability is for dynamic simulation, and it always seems that reservoir engineers need to apply local or regional permeability multipliers for their models to history match. This is often because the porosity–permeability relationship derived from core data has not been calibrated to well-test or mobility data: we tend to underestimate high permeability values and overestimate low values because of the simple algorithms applied.

The arithmetic, geometric and/or harmonic means are most commonly used when trying to provide an average of effective permeability, whether that is for a reservoir interval or a geocellular model. Use of the arithmetic mean returns a higher value than the geometric mean and the harmonic mean gives the lowest result; the effective permeability is likely to be somewhere between this range of values. Because permeability is a vector property, it is possible to capture directional anisotropy by using one average method in one direction (usually laterally) and another method in the other direction (usually vertical); this approach tries to reflect a vertical versus horizontal variation or the K_v/K_h ratio. The geometric mean is normally a good estimate for permeability if it has no spatial correlation and is log-normally distributed. The geometric mean is sensitive to lower values, which will have a greater influence on results.

There are a number of approaches for distributing blocked continuous properties such as porosity; simple two-dimensional zone average mapping produces a smooth result lacking in geology, interpolation between well points in 3D introduces a degree or vertical heterogeneity, but to capture the heterogeneity represented by a reservoir especially supported with seismic data, a stochastic method is required (Figure 10.2).

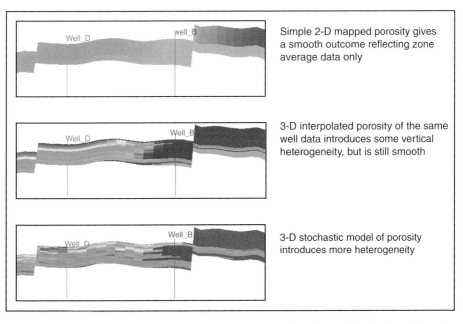

Figure 10.2 Porosity distribution: mapped, interpolated and stochastically distributed showing the increasing heterogeneity in the property. Source: Reproduced courtesy of Roxar Limited. (For a colour version see Plates section).

10.4   Geological issues

In the construction of a geocellular model, there are three geological aspects that must be defined and captured: the structural model, the stratigraphic model and the internal reservoir architecture. To a large extent, these three components are all about the scales of heterogeneity that are being modelled. At the large scale it is the size and nature of interpreted faults and their capacity to compartmentalize a reservoir; the internal layering created by the stratigraphic correlation between wells and the way the geocellular grid is constructed influences the connectivity between reservoir bodies. Within reservoir bodies, the heterogeneity is modelled at a variety of scales, but it is here that petrophysics comes into the mix of variables and uncertainty, usually constrained by small-scale geological factors such as depositional textures, the presence of shale or cement or the impact of fractures on flow in the reservoir.

Therefore, at the largest scale, a reservoir model should capture the structural and stratigraphic framework of a field in 3D and thereby improve the volumetric estimate of hydrocarbons in place in different compartments. Geology is three-dimensional, so we should construct models that reflect this property.

In trying to represent the internal reservoir heterogeneity better, we should always bear in mind that reservoirs are invariably more complex and more uncertain than one thinks; it is seldom that a new well is drilled in a field that comes in exactly as expected and never if drilled on a purely stochastic model! In trying to build a representative model of a field, it is important to achieve the right balance between determinism (what we know or think we know) and probability (what we think we might know, but let the software decide). This is true in all aspects of the process, whether it is the depth conversion uncertainty that can impact GRV by 10–30% or the proportion of net reservoir within closure: a degree of uncertainty needs to be represented by some stochastic property modelling. A word of advice: no model represents the true subsurface character of a reservoir; it represents the available data only and, if properly modelled, a possible realization of the subsurface.

A key component in property modelling is whether to build a facies model to capture the relationship between different types of reservoir body, before distributing the petrophysical data, porosity or permeability. In almost all cases, I think it important that the petrophysical data have been partitioned in a meaningful way to enhance our understanding of the fine-scale distribution of heterogeneity. Where it may not be a fundamental requirement is when post-depositional processes have destroyed the depositional fabric of the reservoir, diagenesis in a carbonate field being the most obvious example. A good facies model guides the internal architecture of a reservoir and the distribution of properties and can explicitly model connectivity in the reservoir, thus impacting the dynamic component of the model (Figure 10.3).

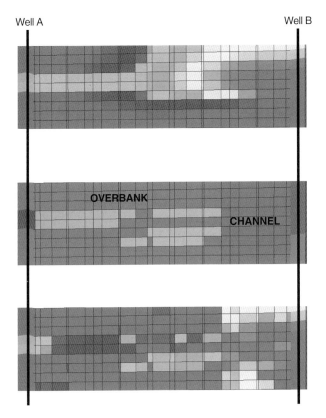

Figure 10.3 Facies-constrained porosity distribution. (a) The interpolated porosity model honours the well data but results in a smooth distribution between the wells; (b, c) a simple threefold scheme of channel, overbank and floodplain facies allows the porosity seen in the wells to be distributed meaningfully, capturing the rapid changes laterally in the model. Source: Reproduced courtesy of Roxar Limited. (For a colour version see Plates section).

Facies modelling falls into two main groups: pixel-based methods and object-based methods. Each method reflects the challenges facing companies at the time of development and the historical preference of the academic institutions where the development work was done. Object modelling evolved primarily in Norway, where oil companies were developing fields with complex fluvio-deltaic environments in the prolific Brent Province of the northern North Sea. They needed to model explicitly connected channel systems in a background of non-reservoir facies. In the United States and France, pixel-based modelling dominated, based on tools such a GSLIB and traditional kriging algorithms developed by the mining industry. One variant on the traditional methods is called multi-point statistics (MPS), where a training image is used to guide the distribution of facies of rock types. Commonly, a mixture of methods can be used to create a geologically sensible model of facies distribution that reflects the conceptual model: the key question to ask is, 'does it look right?'.

The great advantage of a facies model to the petrophysicist is that in property modelling each net contributing facies or rock-type can be represented by the appropriate defining values for porosity or permeability, capturing individual trends, local variability and specific variogram models. This is even more important for saturation modelling, where each facies or rock type can have a unique saturation–height relationship.

10.5   Engineering issues

Dynamic reservoir models have been around for a long time; they were originally built from electrical components where capacitors formed the volume stored and resistors the reservoir heterogeneity to limit flow (Mayer-Gürr, 1976). Today, however, numerical simulators can solve nearly any problem that a reservoir engineer encounters provided that the geological and petrophysical inputs accurately represent the reservoir. The real value of the dynamic model is the ability to provide production profiles and cash flow predictions under different production scenarios, upon which major investment decisions are made, so we had better get the inputs right!

When building a reservoir simulation model, there are many issues that should be considered together with the reservoir geologist. Designing the optimum model grid probably has the greatest impact on the ultimate result. The geologist and the engineer, together with input from the geophysicist and petrophysicist, must design the grid if it is going to capture those parameters that ultimately control flow in the reservoir. Grid design is always a compromise between detail and computational time; the geologist needs to capture large- and small-scale heterogeneities, whereas the engineer requires a grid that is geometrically robust with a sufficient number of cells that correctly handles continuity of flow, without taking a week to make a run.

After grid design, assignment of reservoir properties, or upscaling, provides the greatest challenge. This challenge is directly related to the issues of scale discussed previously; representing data that have been captured at many orders of magnitude different to the modelled property remains the greatest challenge for the reservoir modelling team. Accurate reservoir simulation is only valid when based on a representative geological model and correctly upscaled reservoir properties. Never forget that the key to a good reservoir model is an understanding of the scale of heterogeneity that is important to fluid flow, and this is where petrophysics again come to the fore.

10.6 Volumetrics

Conventional volumetrics calculations to estimate hydrocarbons initially in place (HIIP) require a number of simple input parameters, most of which are in the remit of the petrophysicist; the input parameters are related in the following way:

$$HIIP = \frac{GRV \times NTG \times \phi \left(1 - S_w\right)}{B}$$

where GRV = gross rock volume, NTG = net to gross, ϕ = porosity of net reservoir, $1 - S_w$ = hydrocarbon saturation and B = formation volume factor.

Gross rock volume is the volume of rock between the top reservoir structure and the closing contour or hydrocarbon water contact. In map-based workflows the area of the closure is measured using a planimeter and then the height or reservoir thickness is calculated to give a slab volume; this can be repeated for individual contours to obtain a more accurate result. In a 3D geocellular model, the GRV is calculated more accurately from the total volume of each individual cell between the top structure and the closing contour or hydrocarbon water contact.

Net to gross, as we have previously discussed, is that part of the reservoir containing moveable fluids. In a 3D model, this can be calculated from a facies model or by the application of a series of cut-off values based on an effective porosity and/or permeability. The ratio of NTG to GRV gives the net rock volume (NRV).

Porosity is the capacity of a rock to store fluids and in a 3D model each cell is assigned a porosity value; the pore volume of the model is the total of all cells with an effective porosity and is directly linked to the NRV. Cells with effective porosity may be defined by a facies model or by some cut-off value.

Hydrocarbon saturation is the proportion of the pores filled with hydrocarbon rather than water; the volume of hydrocarbon plus the volume of water will be unity. Only those cells that have been designated as being NRV will be counted in the summation to give the reservoir hydrocarbon volume.

Formation volume factor accounts for the increase in hydrocarbon volume between the reservoir and the surface; this is a function of the change in pressure between reservoir and surface conditions and depends on the fluid description.

10.7 Uncertainty

Every element of reservoir characterization has uncertainty and, as most elements are combined to give an answer, the uncertainty increases at every step in a workflow. Dealing with uncertainty can be a major issue for many people for whom science should provide *the* answer, not *an* answer; geologists should be very comfortable dealing with uncertainty because they are always required to make decisions based on very limited and imprecise data. Investment banks would rather employ a person able to make a decision based on the available data than someone who always requires more information; that way opportunities are lost rather than potential profits are made!

The most fundamental uncertainty is the accurate measurement of a point in 3D space, be this a well surface location or the depth to top reservoir in a model. Most other uncertainties are related to the measurement of properties in the subsurface and the calibration of samples brought to the surface for analysis. Petrophysical measurement is at the heart of these uncertainties, be it correct well to seismic tie or the range of porosity associated with a reservoir rock type. Managing uncertainty is something that comes with experience and, dare I say it, age; understanding the impact of uncertainty on required outcome is even more important, especially if major financial decisions are to be made (Figure 10.4).

Uncertainty comes in two forms: randomness and systematic uncertainty. The former is due to the intrinsic variability in nature and the fact that we never sample all possible values in a property distribution, and the latter is a function of lack of knowledge or an incomplete understanding of the data: this uncertainty can be reduced by making additional measurements, i.e. drill another well!

Looking at the input parameters for the hydrocarbons-in-place calculation, the greatest impact on volume is the range in GRV and the uncertainty associated with the top structure map and the hydrocarbon water contact. The uncertainty of the top structure at every grid node is a function of the seismic horizon

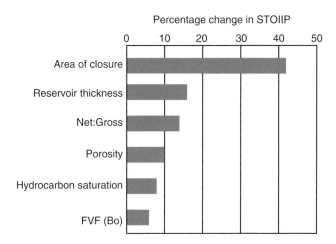

Figure 10.4 Volumes and uncertainty: the percentage change in STOIIP dependent on different petrophysical parameters. The greatest difference comes from the measurement of GRV, due in part to the uncertainty in depth conversion.

interpretation, the well to seismic tie and also the depth conversion, all three of which can have a range of input values. The petrophysicist is able to influence the well to seismic tie by ensuring that the sonic and density logs are correctly edited and depth matched; provide input for velocity modelling, either as instantaneous or interval velocity from sonic logs, checkshots or vertical seismic profiles (VSP); and be able to advise on which wells have the most reliable data and ensure they are correctly distributed across an area of interest: choosing platform wells only for depth conversion will bias the outcome as there will be no variability in the overburden velocity.

Determining what to include as net reservoir is the job of the petrophysicist on a well-by-well basis, but in a 3D model this step should be done with the geologist/reservoir modeller to ensure that the facies model and the petrophysical interpretation are aligned. Both the geologist and petrophysicist will use core analysis data to help define rock types and these should relate to the core-derived facies so that properties such as porosity and permeability are correctly assigned to net and non-net reservoir facies. Where porosity or permeability cut-offs alone are used to define net reservoir, then the subsequent distribution of the properties is divorced from the depositional model, reducing the efficacy of the model, especially in terms of flow connectivity.

A representative 3D model of a reservoir incorporating both deterministic and stochastic data allows the asset team to play the tunes on the uncertainties inherent in all the subsurface disciplines. I hope that having read the previous chapters, it obvious why I refer to 'estimation' rather than 'calculation' of porosity, water saturation and permeability: we can only provide the best approximation of these properties given all the uncertainty in data acquisition, processing and interpretation of the limited data available to us as geophysicists, geologists, reservoir engineers and, of course, petrophysicists.

10.8   Epilog

Finally, my approach to petrophysics may not be that of hard-core field engineers and physicists, but it is a practical way through a myriad of scientific, engineering and technical issues that go to make up our evaluation of oil, gas and water reservoirs. Given the fact that we are seldom able to measure directly the parameters needed for our estimation of porosity, permeability or water saturation, petrophysics allows us to give free rein to our imagination and understanding of the geological and physical interpretation of the subsurface. So, don't just follow the default work-flow or press the quick-look analysis button in your software of chose, but play with the data in a spreadsheet or on the back of an envelope to help you understand how changing a parameter effects the result of your interpretation. And don't forget "garbage in, garbage out"!

Appendix 1
Petrophysical Report

PRELIMINARY PETROPHYSICAL EVALUATION

Operator: COYOTE OIL AND GAS LIMITED
Field: BLACKWELL
Well: WILEY-7
Date: FEBRUARY 2015

Petrophysics: A Practical Guide, First Edition. Steve Cannon.
© 2016 John Wiley & Sons, Ltd. Published 2016 by John Wiley & Sons, Ltd.

Summary

Well Wiley-7 was drilled during November–December 2014 to appraise the Blackwell Group in a new fault block to the SW of the main field. The well reached a total depth of 2160 m in the Blackwell shale, having penetrated 177 m of hydrocarbon-bearing Wiley Formation. The reservoir comprises sands, shales and coal deposits typical of the regional coastal–deltaic depositional model (Table A1.1).

Table A1.1   Wiley-7 header information.

Well	Wiley-7	
Field	Blackwell	
Operator	Coyote	
Block	20/15	
Country	UKLAND	
Elevation	33.2 m	
TD Driller (MD)	2160 m	
Top Interpretation Interval	2044 m	
Base Interpretation Interval	2161 m	
Bit Size	12.25 ins	
Mud Weight	8.0 lb/g	
RM and Temperature	3.00 ohm m	80.0 °C
RMF and Temperature	3.00 ohm m	80.0 °C
RMC and Temperature	3.00 ohm m	80.0 °F
Delta T Shale	100.00 μs/ft	
RhoB Shale	2.55 g/cm^3	
PhiN Shale	35 p.u.	
GR Min/Max	10 GAPI	90 GAPI
BHT	120.00 °F	
R_w and Temperature	0.062 ohm m	80.0 °C
m and n	1.8	1.8

The net-to-gross ratio in the well is ~40% with an average net sand porosity of 13.8% and a net water saturation of ~22%. Permeability has not been estimated for this well. No core was acquired in the well; however, the general field parameters established in Wiley-2 are applied to this interpretation.

Objectives

Well Wiley-7 was drilled to appraise the nature and hydrocarbon potential of the Wiley Formation in a separate compartment of the Blackwell Field. Sufficient field-wide data already exist, making minimum data acquisition a key objective: no cores were cut and the logging suite was simplified from previous wells.

1. **Reservoir Description**

   The Wiley Formation is part of the Blackwell Group; the regional depositional model is of a sequence of coastal–deltaic sediments comprising interbedded sands, shales and coals. The penetrated sequence is 117 m thick and made up of the following units:

   a. *2044–2060 m – Wiley A Sand Member*

   The A Sand forms the uppermost and most productive sand in the main field. It consists of an upward coarsening sequence of deposits typical of deposition in beach and barrier bar environments. The unit can be up to 30 m thick and have a net to gross of 40–100%. In the main field, average porosity is 16% and permeability ranges between 50 and 500 mD. The uppermost part of the unit may be heavily cemented, indicative of the change to deeper marine sedimentation

   b. *2060–2071 m – Wiley Shale Member*

   The Wiley Shale is a field-wide horizon that separates the Wiley A Sand from the underlying Wiley B Sand. The unit comprises 8–15 m of fossiliferous shale and represents a major drowning event.

   c. *2071–2086 m – Wiley B Sand Member*

   The Wiley B Sand is typically 15–20 m thick with a net to gross ratio of 40–100%. It generally consists of up to three distinct upward coars–ening subunits starting with a fine-grained transgressive siltstone–sandstone rich in siderite and shelly debris. Above it are fine-grained sandstones that pass upwards into clean, medium-grained cross-bedded sands typical of a beach or barrier deposit. The sands are generally of good quality although away from the main structure their presence is less predictable.

   d. *2086–2161 m – Wiley Coal Member*

   The lowermost member of the Wiley Formation comprises an intercalated sequence of sandstones, siltstones, shales and coals. The unit varies in thickness across the field and this well has the thinnest penetrated sequence to date. The typical net to gross ratio is 6–20%, with commonly three or four discrete sandstones recorded that are interpreted as poorly connected shoe-string bodies. The sandstones have sharp erosive bases that fine upwards into mottled silty deposits. These interbedded deposits were laid down in a freshwater swamp cut by alluvial channels. At the top of the sequence is a well-developed coal, the Wiley Coal, that represents and extensive, low-lying swamp.

2. **Data Preparation**

   The log data used for the interpretation of the reservoir interval in Wiley-7 are fit for purpose but limited: the log data were acquired as Run 2 of the well. The following logs with value ranges and average value are presented in Table A1.2.

Table A1.2 Available wireline log data from Wiley-7.

Curve name	Curve category	Curve minimum	Curve maximum	Average
MD	Measured depth (m)	2040	2160	–
BS	Bit size (in)	12.25	12.25	–
CALI	Calliper (in)	11.82	19.91	13.15
GR	Gamma ray (API)	0.80	125	54.0
RHOB	Bulk density (g/cm³)	1.47	2.71	2.35
DRHO	Density correction	−0.12	0.26	0.08
DT	Sonic (µs/ft)	51.5	124.4	77.1
NPHI	Neutron porosity (p.u.)	2.17	56.3	25.0
ILD	Deep induction (ohm m)	3.45	73.71	11.34
LLD	Laterolog deep (ohm m)	2.922	158.60	16.565
LLS	Laterolog shallow (ohm m)	2.48	103.63	14.146
MSFL	Micro resistivity (ohm m)	0.517	214.37	15.481

Overall log quality is acceptable, although there are significant borehole sections where the density correction is unacceptable and the sonic porosity has been substituted. A minimal amount of editing of raw data has been undertaken and there are no significant depth shifts required. Figure A1.1 presents a composite of the primary raw data input for the analysis.

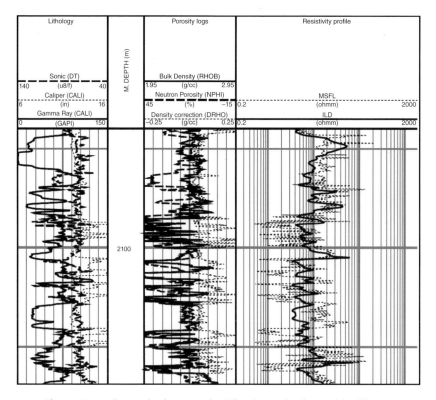

Figure A1.1 Composite log over the Wiley Formation interval in Wiley-7.

2.1 *Environmental corrections*

Formation temperature was taken from the density–neutron combination log run, as this was the maximum value measured and coincident with the regional temperature gradient. The mean surface temperature was taken as 14.7 °C, typical for this location and season.

The gamma ray, density, neutron and resistivity logs were all corrected for borehole effects using the appropriate correction algorithms provided by the service company and built into the software. Invasion corrections were also made using mud resistivity data and a formation salinity of 35,000 ppm, equivalent to a formation resistivity of 0.062 ohm m at 80 °C.

3. Log analysis

Log analysis has been done in the Terrasciences T-Log Version 2 software. Both 'quick look' and detailed interpretation workflows have been applied; however, only the latter is reported here. The Wiley Formation has been interpreted as a single unit; it may be appropriate at a later stage to subdivide the interval should more core data become available.

3.1 *Shale volume estimation*

V_{sh} has been estimated from the gamma ray log using a simple min/max approach derived from a histogram display of values over the interpretation interval. The minimum, 'clean sand' value is taken as 10 API and the maximum 'shale' value as 90 API representing the bulk of the data and ignoring insignificant 'tails' (Figure A1.2).

3.2 *Archie parameters*

The field-wide Archie parameters have been established from core data and are used in this analysis: $a = 1, m = 1.8, n = 1.8$.

3.3 *Formation water resistivity*

Formation water resistivity has been obtained from water samples taken by RFT in the main field, as there is no water-bearing sand penetrated by the well. Formation water salinity is 35,000 ppm chlorides, equivalent to a formation resistivity of 0.062 ohm m at 80 °C.

3.4 *Matrix, shale and fluid properties*

Matrix properties were estimated from the neutron–density crossplot and from core data taken from well Wiley-2. Grain density is 2.65 g/cm³ in the sandstone.

Shale characteristics were reviewed by examining plots of V_{sh} against density, sonic and neutron logs to estimate limiting values where $V_{sh} = 1$. Shale points for each log are given as follows:

$RHOB_{sh} = 2.55$ g/cm³
$DT_{sh} = 100$ μs/ft
$PhiN_{sh} = 35$ p.u.

Shale corrections were automatically made in the porosity calculation.

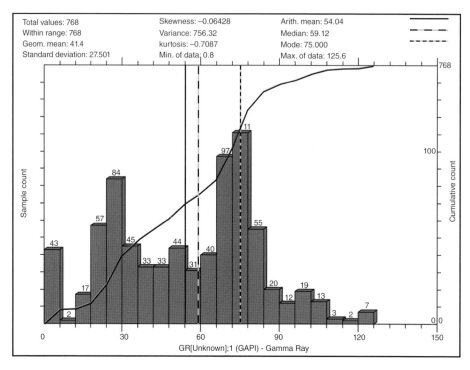

Figure A1.2   Gamma ray histogram plot for Wiley Formation interval in Wiley-7.

Hydrocarbons are present in the well and the logs should be corrected before using them to estimate porosity. Apparent fluid density was generated by regression analysis of overburden-corrected core porosity and log-measured density over the corresponding reservoir interval, in this case a comparison between well Wiley-2 and the subject well. The regression was confined to intervals of clean sand only using the following constraints:

$$\text{Calliper-bit size} < 2 \text{ in and } V_{sh}\left(\text{GR}\right) < 0.15$$

The regression should be carried out for both oil-leg and water-leg separately, but in this case there is no water-bearing sand penetrated. The apparent oil density is 0.81 g/cm³ and the default water density for the field is 1.03 g/cm³. The hydrocarbon corrections were carried out automatically in the porosity calculation.

3.5   *Porosity estimation*

Initially two porosity estimations were made using the density log alone and the neutron–density combination; this was because of observed washouts in the well. A third, sonic porosity, was calculated for intervals where neither density alone nor neutron–density combination returned a sensible solution. The washed-out sections were primarily in the shales and coals of the Wiley Coal Member.

Density porosity (ϕ_{DEN}) was estimated using the shale- and hydrocarbon-corrected density log values in the following equation:

$$\phi_{DEN} = \frac{\rho_{matrix} - \rho_{corrected}}{\rho_{matrix} - \rho_{fluid}}$$

Neutron–density porosity (ϕ_{N-D}) was evaluated from hydrocarbon- and shale-corrected density and neutron logs using a mathematical solution to the conventional neutron–density cross-plot in Terrastation using the fluid and matrix parameters described earlier.

Porosity Selection Criteria

A comparison of the different methods showed broad conformance although the sonic porosity was generally more conservative. Where the borehole constraint (calliper bit size <2 in) was met, the neutron–density porosity was used; elsewhere the sonic porosity was substituted. This approach gives a robust measure of total porosity in the reservoir interval. Effective porosity can be calculated from the total porosity and the clay-bound water saturation:

$$\phi_E = (1 - S_{wc})\phi_T$$

and S_{wc} is the saturation of clay-bound water calculated from

$$S_{wc} = \frac{\phi_{Tsh}V_{sh}}{\phi_T}$$

or

$$\phi_E = \phi_T - \phi_{sh}V_{sh}$$

3.6 *Saturation estimation*

The water saturation was evaluated using both the Archie equation and the modified Simandoux total shale relationship.

Archie:

$$S_w = \left(\frac{aR_w}{R_t\phi^m}\right)^{\frac{1}{n}}$$

Simandoux:

$$S_w = \left\{ \frac{aR_w(1-V_{sh})}{\phi^m R_t} + \left[\frac{V_{sh}aR_w(1-V_{sh})}{2\phi^m R_{sh}}\right]^2 \right\}^{\frac{1}{2}} - \frac{V_{sh}aR_w(1-V_{sh})}{2\phi^m R_{sh}}$$

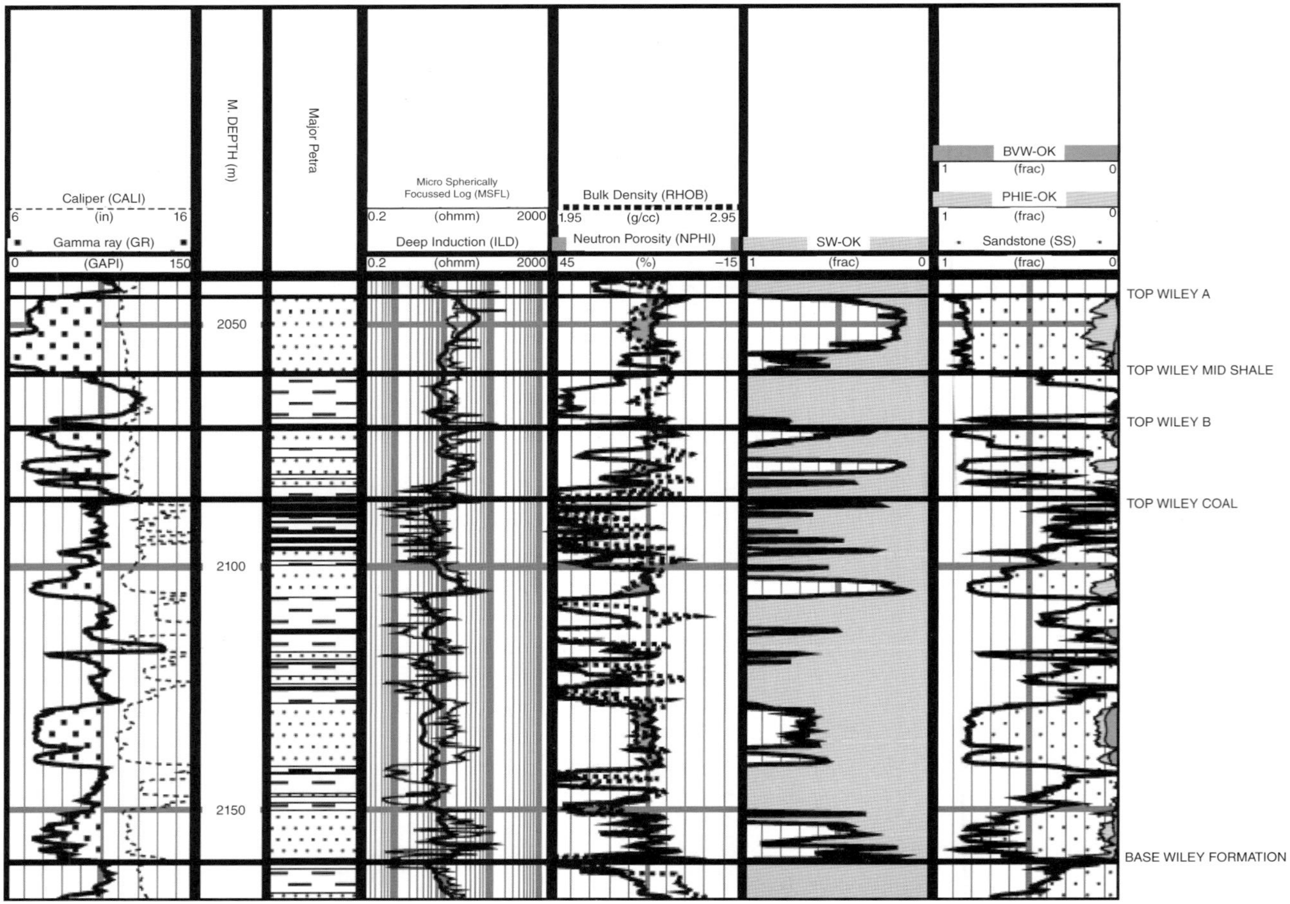

Figure A1.3 CPI of Wiley Formation interval in Wiley-7.

Table A1.3 Final petrophysical interpretation of the Wiley Formation interval in Wiley-7.

Unit	Interval (m)	Net reservoir (m)	Net:gross	Net pay	ϕ_{total}	V_{sh}	S_w
Wiley A Sand	2044–2060	15.39	0.953	8.99	0.144	0.07	0.26
Wiley Mid-Shale	2060–2071	Non-reservoir					
Wiley B Sand	2071–2086	5.57	0.364	2.59	0.126	0.11	0.26
Wiley Coal Member	2086–2161	15.09	0.200	2.59	0.130	0.18	0.26
Wiley Formation	2044–2161	36.0	0.31	14.2	0.138	0.10	0.26

Values for the Archie parameters and formation resistivity have been described earlier and can be used directly in the software algorithms.

A comparison of the Archie and Simandoux results shows a strong correspondence in the clean sand intervals, as expected. The Archie equation overestimates water saturation in the shaly intervals; therefore, the Simandoux equation was used throughout.

A bulk water volume (BVW) was calculated from the interpreted water saturation and porosity primarily for display purposes.

4. **Summary of results**

The final results are presented as a computer-processed interpretation (CPI) (Figure A1.3). The following cut-offs were applied to establish the net reservoir rock in the interval:

Volume of shale <0.30
Water saturation <0.40
Total porosity   <0.25

The net reservoir in the total interval (117 m) is 36 m, giving a net to gross ratio of 31%: net pay, based on the cut-offs, is 14.2 m. Within the net reservoir, mean porosity is 13.8%, mean V_{sh} is 1% and mean S_w is 26%. Table A1.3 shows the same information for each interval.

Appendix 2
Data Collection
and Management

Data management is probably the most important part of any petrophysical project. About 50% of the project schedule is spent on preparing data for loading and checking for inconsistencies. The quality of input data is the essential element of a project; if there are any inconsistencies in the input data, they will show up in the end result and at every intermediate step.

Available input data

First define what data are available and what can be achieved with them, before starting the project. A data delivery schedule should be established for any additional data or interpretations that may be included in the project. Projects often veer off track when new data are supplied after a milestone has been passed in the workflow. Often it is worthwhile agreeing a cut-off date for any additional data.

Quality

The quality of the input data used should be reviewed at the start of the project to judge what can be done to improve or replace them. Inconsistencies in the data can cause numerous problems at every step of the modelling process.

Petrophysics: A Practical Guide, First Edition. Steve Cannon.
© 2016 John Wiley & Sons, Ltd. Published 2016 by John Wiley & Sons, Ltd.

Database

Most companies possess a corporate database for data storage; Openworks and Geoframe are the most common in use. The management of the corporate database is not a subject for the project team; however, there is a requirement to access the database for input data and often to return completed interpretations or results to the database. It is recommended that one member of the project team is tasked with organization and management of both the project data store and the relationship with the corporate database.

Data types

A surprisingly small amount of data is required to build a simple, representative petrophysical model, but to capture the range of uncertainties in a reservoir more information will often be required than is available. Rarely there may even be too much data and an overcomplicated model can result that ultimately may have little practical value.

The primary data types used for reservoir modelling are

- **seismic**-derived interpretations and processed volumes
- **well** data, including deviation data, cores, logs and pressures
- **dynamic** data from well test and production.

Seismic data

The task of the seismic interpreter is to provide key mappable horizons and faults from which the structural framework of the geological model can be constructed. The horizons and fault input data are interpreted in time and subsequently depth converted using a velocity model. Since the interpretation is done in two dimensions, it is recommended to visualize the data in 3D and determine if there are any inconsistencies.

Velocity model
The velocity model is generally stored in the project data store either as a grid or as a cube. The data can be loaded into the project if depth conversion is to be carried out as part of the modelling exercise. Alternatively, depth-converted surfaces and fault information are provided.

Well data

Well data are generally point source data from a non-regular 'grid' over the area of interest. As such, the data represent a very small investigation volume for the reservoir that is to be modelled, but they are also the 'hardest' data available to the project team. Consistency in all aspects of these data is crucial, especially in projects where there may be many wells to include; consistent well names and basic datum information should be checked thoroughly prior to loading. Different databases have different ways of storing and exporting these fundamental data and an experienced operator should be involved in extracting the data efficiently.

Wellbore path

A single quality-controlled collection of the current valid survey data set should be available. Uniqueness of wellbore path data is essential. If editing of the data is necessary, old versions should be removed or flagged in a way that will avoid later erroneous use.

Wellbore path data are computed from suitable corrected direction survey data and are stored in the database. Wellbore path data should consist of (x, y, z) locations and a surface location. The data should be regularly sampled along the wellbore (i.e., regularly sampled in MD) with an interval of less than 1 m (usually 0.15 m).

Composite log

It is important to ensure that a complete set of quality-controlled composite data is available for spatial modelling and property analyses. Table A2.1 gives a listing of curves and specific log names that should be defined in the modelling database.

Logs should be pre-processed for modelling purposes and a set of curves assembled either in the project data store or in a separate petrophysical database. These data will form the 'basic input data' for reservoir modelling.

Some or all, of the following processing may be advantageous:

- All logs should be resampled to a common regular sample interval.
- For multi-well analysis, environmental correction and normalization of logs may be beneficial.
- Bad data, such as sonic log spikes or washed-out intervals, should be excluded.
- Removal of shoulder effects (buffering) with a masking method can be a useful precursor to blocking of well data.

Computer-processed interpretation (CPI) logs

Table A2.2 gives a listing of CPI curves that should be available in the modelling database, together with naming conventions. Many of the curves may not be available in any current corporate database and will need to be created and stored in the project data store if required. All curves must be available in a regular resampled form.

Table A2.1 Common mnemonics.

Mnemonic	LIS mnemonic	Description	Measurement units
AC_COMP	ACCP	Compressional sonic log	Delta T, μs/ft
AC_SHEAR	ACSH	Shear sonic log	Delta T, μs/ft/ft
AC_STONE	ACST	Stoneley wave	Delta T, μs/ft
BIT_SIZE	BS	Drill bit diameter	in
CALI	CALI	Calliper; borehole diameter	in
CMFF	CMFF	Free fluid index from CMR	
CMRP_3MS	CMR3	Nuclear magnetic resonance	Relaxation time, s
DEN	DEN	Bulk density (RhoB)	g/cm^3
DEN_CRN	DENC	Bulk density correction	g/cm^3
FTEMP	FTEM	Formation temperature	°C or °F
GR	GR	Natural gamma ray	API
GR_KTH	GRKT	Potassium/thorium ratio	
K	K	Permeability	mD
KTIM	KTIM	Permeability using Timur	mD
NEUT	NEU	Neutron porosity	Hydrogen index/p.u.
PEF	PEF	Photoelectric absorption effect	barns/electron
RES_DEP	RDEP	Deep resistivity	ohms/m (Ω/m)
RES_MED	RMED	Medium resistivity	Ω/m
RES_MIC	RMIC	Micro-resistivity	Ω/m
RES_SHA	RSHA	Shallow resistivity	Ω/m
RT	RT	Formation (true) resistivity	Ω/m
RXO	RXO	Invaded zone resistivity	Ω/m
SGR	NGS	Spectral gamma ray	API
SP	SP	Spontaneous (self) potential	Direct current mV
TCMR	TCMR	Relaxation time from CMR	s
TEN	TEN	Tool tension	lb
TH	TH	Thorium from SGR	p/p
URAN	U	Uranium SGR	

Core descriptions

Core descriptions are used for geological analysis and the following types should be available for review:

- Detailed 1:50 sedimentological core description.
- High-resolution structural core description usually at a scale of 1:40 for use in fracture modelling.
- Log calibrated 1:200 generalized description to tie core and log petrophysical properties such as porosity and permeability.

It is important to shift the description to loggers' depth to achieve correspondence between the core description measured depth and the

Table A2.2 Common CPI mnemonics.

CPI	Abbreviation	Description	Units/Boolean
BADHOLE	BDHL	Over gauge hole	Flag
CALCITE	CALC	Calcite	Flag
COAL	COAL	Coal	Flag
EFAC	EFAC	Electrofacies	Flag
LIMESTONE	LIME	Limestone	Flag
PAY	PFLG	Pay flag	Flag
PERM	LPRM	Permeability	mD
PERM_NET	LKNT	Net permeability	mD
POR_EFF	PORE	Effective porosity	v/v
POR_EFF_NET	PREN	Net effective porosity	v/v
POR_TOT	PORT	Total porosity	v/v
RESERVOIR	RFLG	Reservoir flag	Flag
SAND	SAND	Sand	Flag
SW_EFF	SWE	Effective water saturation	v/v
SW_EFF_NET	SWEN	Net effective water saturation	v/v
SW_IRR_EFF	SWIE	Effective irreducible SW	v/v
SW_IRR_TOT	SWIT	Total irreducible SW	v/v
SW_TOT	SWT	Total water saturation	v/v
SWE_MOD	SWEC	Modified effective SW	v/v
SWT_MOD	SWTC	Modified total SW	v/v
SXO_EFF	SXOE	Effective SW in invaded zone	v/v
SXO_TOT	SXOT	Total SW in invaded zone	v/v
TAR	TAR	TAR – dead oil	Flag
VANHYD	VANH	Volume of anhydrite	v/v
VBASE	VBAS	Volume of basalt	v/v
VCALCITE	VCAL	Volume of calcite	v/v
VCBW	VCBW	Volume of clay-bound water	v/v
VCHALK	VCHK	Volume of chalk	v/v
VCHERT	VCHT	Volume of chert	v/v
VCLAY	VCL	Volume of clay (diagenetic)	v/v
VCOAL	VCOL	Volume of coal	v/v
VDOLO	VDOL	Volume of dolomite	v/v
VHALITE	VHAL	Volume of halite	v/v
VLIME	VLIM	Volume of limestone	v/v
VMARL	VMRL	Volume of marl	v/v
VOOZE	VOOZ	Volume of ooze	v/v
VSAND	VSND	Volume of sand (quartz)	v/v
VSHALE	VSH	Volume of shale	v/v
VSILT	VSLT	Volume of silt	v/v
VTAR	VTAR	Volume of tar	v/v
VTUFF	VTUF	Volume of tuff	v/v
VVOLC	VVOL	Volume of volcanic	v/v
VWAT_EFF	7VWE	Volume of moveable water	v/v
VWAT_TOT	VWT	Volume of total water	v/v
VXOWAT	VXOW	Volume of water in invaded zone	v/v

Table A2.3 Common core analysis mnemonics.

Core plugs	Abbreviation	Description	Measurement units
CORENUMBER	CORE	Sequential core number	–
CPOR	CPOR	Helium porosity	% or fraction
CPORF	PORF	Fluid-filled porosity	% or fraction
CPOROB	PORB	Overburden-corrected porosity	% or fraction
CPORV	PORV	Pore volume	v/v
CPORVOB	POVB	Overburden-corrected pore volume	v/v
CSG	CSG	Gas saturation	% or fraction of PV
CSO	CSO	Oil saturation	% or fraction of PV
CSW	CSW	Water saturation	% or fraction of PV
GRDEN	GRDE	Grain density	g/cm^3
KHKL	KHKL	Klinkenberg-corrected horizontal permeability	mD
KHL	KHL	Horizontal liquid permeability	mD
KHLOB	KHLB	OB-corrected liquid horizontal permeability	mD
KHOB	KHOB	OB-corrected horizontal permeability	mD
KHOR	KHOR	Horizontal permeability	mD
KVER	KVER	Vertical permeability	mD
KVKL	KVKL	Klinkenberg-corrected vertical permeability	mD
KVL	KVL	Vertical liquid permeability	mD
KVLOB	KVLB	OB-corrected vertical liquid permeability	mD
KVOB	KVOB	OB-corrected vertical permeability	mD
LITH	LITH	Lithology	Identifier
MDEPTH_RCA	MDEP	Measured depth routine core analysis level	ft or m
SAMPDIAM	PLUD	Sample/plug diameter	cm or in
SMID	SMID	Sample/plug identifier	Identifier

petrophysical logs. Be on the lookout for incomplete core recovery and misplaced sections.

Core photographs
Core photographs should be shifted to logger's depth along with the core itself. Core photographs become the permanent data record as the core material may be accessed repeatedly and often disturbed through handling. It is wise to make core/log shift curves and store these in the relevant database. These can then be used with all the core-based data in a consistent fashion.

Core plug data

A unique set of measurement results from core plug data (routine core analysis) should be assembled in the appropriate database for petrophysical modelling. For data to be considered 'final', they should be overburden corrected (porosity and permeability) and permeability corrected for gas slippage, the Klinkenberg correction.

The plug number is useful for comparison with core photographs.

Uncorrected measurements should not be stored in the interpretation dataset as this can be a source of confusion. Data should first be shifted according to a master composite log and an overburden correction of data is necessary.

Core measurements, unlike log measurements, are not sampled at regular increments, hence they should be considered as discrete rather than continuous data to avoid interpolation between points during data import. For use in data analysis and subsequent property modelling, core plug measurements must have been shifted to the closest depth increment in the corresponding CPI and/or composite data.

Reservoir zonation

There are likely to be a number of different reservoir zonations stored in company archives often generated by different disciplines; the geologist's sequence stratigraphic approach compared with the petrophysicist's flow zonation. A guiding principle should be to integrate the coarsest scale of zonation, usually the seismic interpretation, with a sequence stratigraphic breakdown of intervening field-wide horizons; any debateable correlations should be excluded until their significance is understood. Other zonations can be stored in the project database, but these should be clearly defined as to origin and specific use; for instance, a flow-based zonation may be compared with a geological one to determine areas of commonality.

Pressure data

Pressure data (see Table A2.4) should be stored in the database. The data are often incomplete and will need to be carefully reviewed for validity.

Fluid data

Oil and gas fluid data are required to evaluate the properties of produced fluids at reservoir conditions, in production tubing, in process facilities and in pipeline transportation. The key PVT (pressure–volume–temperature) properties to be determined for a reservoir fluid include the following:

- Original reservoir fluid composition(s).
- Saturation pressure at reservoir temperature.
- Oil and gas densities.
- Oil and gas viscosities.

Table A2.4 Pressure measurement mnemonics.

Mnemonic	LIS	Description	Measurement units
COMMENT	REM	General remarks on quality	
FPRESS	PRES	Formation pressure	psia; bara
FTEMP_FT	TEMP	Formation temperature from the FMT	°C or °F
MDEPTH_FT	MDEP	Measured depth with FMT	ft or m
MOBILITY	MOBL	Mobility calculated from pressure profile	mD/cP
RUN_NO	RUN	Logging tool run	
FT_SYMBOL			
TEST_NO		Test sample number	
TVDSS	TVDS	True vertical depth sub-sea	ft or m

- Gas solubility in reservoir oil (GOR, R_s).
- Liquid content of a reservoir gas.
- Shrinkage (volume) factors (B_o, B_g, B_w) of oil, gas and water from reservoir to surface conditions.
- Compositional variation with depth.
- Equilibrium phase compositions.

Reservoir fluid volumes are generally reported in stock-tank volumes and the shrinkage factor is therefore a critical property. It should be noted that this property is related to the actual process by which the reservoir fluid is established. Usually shrinkage factors are calculated by equation of state (EOS) simulations. Experimental data are used indirectly, to tune the EOS parameters.

Well-test data

Well-test data can be used to determine effective permeability and are divided into the following types:

- Transient well-test (DST) raw data: rates and pressures.
- Transient well tests: perforation intervals and main interpretation data.
- Transient well-test interpretations: permeability–thickness, skin, boundaries.
- Production log (PLT) interpretations: oil, gas and water rates in the well, plus pressure distribution.

Important specialist data

Special seismic cubes and seismic test lines
These could include coherence, inverted, 4D and pre-stack depth migrated cubes. These cubes should be stored in the project database alongside the standard reflection seismic data for easy visualization and interpretation.

Table A2.5   SCAL mnemonics.

Measurement	Type	Description
Archie a	Electrical	Porosity coefficient
Archie m	Electrical	Cementation exponent
Archie n	Electrical	Resistivity index
Q_v	Cation exchange	
P_c	MICP, porous plate	Capillary pressure
WET	AMOTT/USBM	Wettability
K_{rel}	Steady/unsteady state	Relative permeability

SCAL data

Special core analysis data should be regularly collected for defining petro-physical interpretation parameters and for dynamic measurements. Routinely collected data include those listed in Table A2.5.

These values are all collected and used in the petrophysical assessment of the reservoir and later in dynamic modelling. If using advanced 3D saturation modelling tools in the static model, such as Geo2Flow, these data will also be required.

Borehole image logs and interpretations

Because of their size, raw and processed borehole image data are not gener-ally available online and are stored on tape or CD as part of a service company report. Ideally, the interpretations (depth, dip, dip azimuth, dip type) should be stored in the project database.

Appendix 3
Oilfield Glossary

Term	Application
Abnormal pressure	A subsurface condition in which the pore pressure of a geological formation exceeds or is less than the expected or normal, formation pressure. When impermeable rocks such as shales are compacted rapidly, their pore fluids cannot always escape and must then support the total overlying rock column, leading to abnormally high formation pressures. Excess pressure, called overpressure or geopressure, can cause a well to blow out or become uncontrollable during drilling. Severe underpressure can cause the drillpipe to stick to the underpressured formation
Annular pressure	Fluid pressure in the annulus between tubing and casing or between two strings of casing
Annular velocity	The speed at which mud or cement moves in the annulus; important to monitor to ensure that the hole is being cleaned of cuttings during drilling and to avoid erosion of the borehole
Annulus	The space between two concentric objects such as a borehole and drillstring or casing
American Institute of Petroleum (API)	A trade association founded in 1919 with offices in Washington, DC, USA. The API is sponsored by the oil and gas industry and is recognized worldwide
Azimuth (AZ), azimuthal	The compass direction of a directional survey or of the wellbore as planned or measured by a directional survey. The azimuth is usually specified in degrees with respect to the geographic or magnetic north pole. In well logging: pertaining to being focused in one direction. An azimuthal or azimuthally focused, measurement has one or more directions perpendicular to the surface of a logging tool from which it receives most of its signal

(Continued)

Petrophysics: A Practical Guide, First Edition. Steve Cannon.
© 2016 John Wiley & Sons, Ltd. Published 2016 by John Wiley & Sons, Ltd.

(Continued)

Term	Application
Barite	Weighting material with a specific gravity of 4.37 used to increase the apparent density of a liquid drilling fluid system. Barite ($BaSO_4$) is the most common weighting agent used today. It is a mined material ground to an API specification such that particle sizes are predominantly in the 3–74 μm range
Bottom-hole assembly (BHA)	The lower portion of the drillstring, consisting of (from the bottom up in a vertical well) the bit, bit sub, a mud motor (in certain cases), stabilizers, drill collars, heavy-weight drillpipe, jarring devices ('jars') and crossovers for various threadforms. The bottom-hole assembly must provide force for the bit to break the rock (weight on bit), survive a hostile mechanical environment and provide the driller with directional control of the well
Bit	The tool used to crush or cut rock. Everything on a drilling rig directly or indirectly assists the bit in crushing or cutting the rock. The bit is on the bottom of the drillstring and must be changed when it becomes excessively dull or stops making progress. Most bits work by scraping or crushing the rock, or both, usually as part of a rotational motion
Bit nozzle	The part of the bit that includes a hole or opening for drilling fluid to exit. The hole is usually small (diameter around 0.25 in) and the pressure of the fluid inside the bit is usually high, leading to a high exit velocity through the nozzles that creates a high-velocity jet below the nozzles. The sizes of the nozzles are usually measured in 1/32 in
Bleed off	To equalize or relieve pressure from a vessel or system. At the conclusion of high-pressure tests or treatments, the pressure within the treatment lines and associated systems must be bled off safely to enable subsequent phases of the operation to continue
Blowout	An uncontrolled flow of reservoir fluids into the wellbore and sometimes catastrophically to the surface. A blowout may consist of salt water, oil, gas or a mixture of these
Blowout preventer (BOP)	A large valve at the top of a well that may be closed if the drilling crew loses control of formation fluids. By closing this valve (usually operated remotely via hydraulic actuators), the drilling crew usually regains control of the reservoir and procedures can then be initiated to increase the mud density until it is possible to open the BOP and retain pressure control of the formation
Borehole	The wellbore itself, including the open-hole or uncased portion of the well. Borehole may refer to the inside diameter of the wellbore wall, the rock face that bounds the drilled hole

Term	Application
Bottom-hole pressure (BHP)	The pressure at the bottom of a well, usually measure in bars. In a static, fluid-filled borehole BHP = ρgh, where ρ is the density of the fluid, g is the gravitational constant and h is the depth of the well or height of the fluid column
Bottoms-up	1. Pertaining to the mud and cuttings that are calculated or measured to come from the bottom of the hole since the start of circulation 2. The sample obtained at the bottoms-up time or a volume of fluid to pump, as in 'pump bottoms-up before drilling ahead'
Calliper	Measures borehole diameter and rugosity
Cased hole, casing	The portion of the wellbore that has had metal casing placed and cemented to protect the open hole from fluids, pressures, wellbore stability problems or a combination of these. Large-diameter pipe lowered into an open hole and cemented in place
Circulation system	The complete path that the drilling fluid travels from the rig pumps through the kelly and drill-pipe to the bit and back to the shakers
Circulation time	The elapsed time for mud to circulate from the suction pit, down the wellbore and back to surface. Circulation time allows the mud engineer to catch 'in' and 'out' samples that accurately represent the same element of mud in a circulating system. Circulation time is calculated from the estimated borehole volume and pump rate and can be checked by using tracers such as calcium carbide
Completion	The hardware used to optimize the production of hydrocarbons from the well. This may range from a simple a packer on tubing above an open-hole completion ('barefoot' completion), to a system of mechanical filtering elements outside of perforated pipe, to a fully automated measurement and control system that optimizes reservoir economics without human intervention (an 'intelligent' completion)
Core, coring	To deepen the wellbore by way of collecting a cylindrical sample. A core bit is used to accomplish this, in conjunction with a core barrel and core catcher. The bit is usually a drag bit fitted with either PDC or natural diamond cutting structures, but the core bit is unusual in that it has a hole in its centre
Cuttings	Small pieces of rock that break away due to the action of the bit teeth. Cuttings are screened out of the liquid mud system at the shale shakers and are monitored for composition, size, shape, colour, texture and hydrocarbon content
Depth reference, datum	The point in a well from which depth is measured. It is typically the top of the kelly bushing or the level of the rig floor on the rig used to drill the well. The depth measured from that point is the measured depth (MD) for the well

(Continued)

(Continued)

Term	Application
Derrick	The structure used to support the crown blocks and the drillstring of a drilling rig
Deviated well	A wellbore that is not vertical. The term usually indicates a wellbore intentionally drilled away from vertical at a geological target
Driller's depth	Driller's depth is the first depth measurement of a wellbore and is taken from the rotary table level on the rig floor. It is calculated by adding the length of the BHA plus the drill-pipe
Drilling break	A sudden increase in the rate of penetration during drilling. When this increase is significant, it may indicate a formation change, a change in the pore pressure of the formation fluids, or both
Equivalent circulating density (ECD)	While circulating, the bottom-hole pressure increases by the amount of fluid friction in the annulus. This pressure appears as an apparent increase in mud density, the ECD
Filter cake, mudcake	The residue deposited on a permeable interval when drilling fluid is forced against the borehole wall under a pressure. Filtrate is the liquid that passes through the formation, leaving the cake on the wall
Fish, fishing	Anything dropped or left in a wellbore. The fish may consist of junk metal, a hand tool, a length of drill-pipe or drill collars or an expensive MWD and directional drilling package. Fishing is to attempt to retrieve a fish from a wellbore
Flushed zone	The volume close to the borehole wall in which all of the moveable fluids have been displaced by mud filtrate. The flushed zone contains filtrate and the remaining hydrocarbons, the percentage of the former being the flushed zone saturation, S_{xo}. Also known as the invaded zone
Formation exposure time	The time that has elapsed between the bit first penetrating a formation and a log being recorded opposite the formation
Formation resistivity, R_t	True resistivity of rock plus fluids
Formation volume factor	Ratio of oil volume at reservoir and surface conditions
Fracture gradient	The pressure needed to induce fractures in formation at a given depth
Geopressure gradient	The change in pore pressure per unit depth, typically in units of pounds per square inch or per foot (psi/ft) or kilopascals per metre (kPa/m). The geopressure gradient might be described as high or low if it deviates from the normal hydrostatic pressure gradient of 0.433 psi/ft (9.8 kPa/m)
Geothermal gradient	The natural increase in temperature with depth in the Earth. Temperature gradients vary widely over the Earth, sometimes increasing dramatically around volcanic areas. The down-hole temperature can be calculated by adding the surface temperature to the product of the depth and the geothermal gradient

Term	Application
Gross rock volume	Volume of rock above a fixed datum in a three dimensional model
Gas/stock tank oil initially in place	Hydrocarbons in place at time of discovery (GIIP/STOIIP). Usually infers the volume at surface conditions
Hook load	The total force pulling down on the hook assembly of the drilling rig. This includes the weight of the drillstring in air, the drill collars and any ancillary equipment, reduced by any force that tends to reduce that weight, such as the buoyancy of the drilling fluid
Inclination	The deviation from vertical, irrespective of compass direction, expressed in degrees. Inclination is measured initially with a pendulum mechanism and confirmed with MWD accelerometers or gyroscopes
Injection well	A well in which fluids are injected rather than produced, the primary objective typically being to maintain reservoir pressure. Water-injection wells are common offshore, where filtered and treated seawater is injected into a lower water-bearing section of the reservoir
Junk	Anything in the wellbore that is not supposed to be there. The term is usually reserved for small pieces of steel such as hand tools, small parts, bit nozzles, pieces of bits or other downhole tools and remnants of milling operations
Kelly, kelly bushing	The kelly is used to transmit rotary motion from the rotary table or kelly bushing to the drillstring, while allowing the drillstring to be lowered or raised during rotation. The kelly goes through the kelly bushing, which is driven by the rotary table. The kelly bushing has an inside profile matching the kelly's outside profile (either square or hexagonal), but with slightly larger dimensions so that the kelly can move freely up and down inside
Lag time	The time taken for cuttings to reach the surface. The term is also used in place of cycle time
Leak off test (LOT)	A test to determine the strength or fracture pressure of the open formation, usually conducted immediately after drilling below a new casing shoe. During the test, the well is shut in and fluid is pumped into the wellbore to increase gradually the pressure that the formation experiences. At some pressure, fluid will enter the formation or leak off, either moving through permeable paths in the rock or by creating a space by fracturing the rock. The results of the leakoff test dictate the maximum pressure or mud weight that may be applied to the well during drilling operations
Logging/measurement while drilling	Real-time telemetry and sensor measurements (LWD/MWD). The measurement of formation properties during drilling of the hole or shortly thereafter, through the use of tools integrated into the bottom-hole assembly. LWD measurement ensures that some measurement of the subsurface is captured in the event that wireline operations are not possible

(Continued)

(Continued)

Term	Application
Mud, mud weight	Drilling fluid; density of drilling fluid, usually in pounds per gallon or specific gravity
Mud logger	The person responsible for collecting cuttings samples for geological description and storage, analysing cuttings, gas measurements and analysis and creating a lithological log (mudlog). Often holds a degree in geology or a related discipline
Normal pressure	The pore pressure of rocks that is considered normal where the change in pressure per unit of depth is equivalent to hydrostatic pressure. The normal hydrostatic pressure gradient for fresh water is 0.433 psi/ft and 0.465 psi/ft for water with 100,000 ppm total dissolved solids
Net-to-gross ratio	Ratio of reservoir/pay to non-reservoir
Offset well	An existing penetration close to a proposed well that provides information for planning or interpreting the new well
Oil-based mud (OBM)	A mud in which the external phase is a product obtained from oil, such as diesel oil or mineral oil. More generally, a mud system that has any type of non-aqueous fluid as the external phase including synthetic mixtures
Open hole	The uncased portion of a well. All wells, at least when first drilled, have open-hole sections, prior to running casing. The well planner must consider how the drilled rock will react to drilling fluids, pressures and mechanical operations over time
Overpressure, overbalance	The amount of pressure in the wellbore that exceeds the pressure of fluids in the formation. This excess pressure is needed to prevent reservoir fluids (oil, gas, water) from entering the wellbore
Perforate	To create holes in the casing or liner to achieve efficient communication between the reservoir and the wellbore allowing production of hydrocarbons. A perforating gun assembly can be deployed on wireline, tubing or coiled tubing with the appropriate configuration of shaped explosive charges
Pull-out-of-hole (POOH)	To recover the drillstring from the wellbore; to trip out
Rate of penetration (ROP)	Drilling rate in feet or metres per hour
Resistivity of formation water (Rw)	Function of water salinity
Rotary steerable system (RSS)	A tool designed to drill directionally with continuous rotation from the surface, eliminating the need to slide a steerable motor
Rotary table	The revolving or spinning section of the drill floor that provides power to turn the drillstring in a clockwise direction (as viewed from above). When the drillstring is rotating, the drilling crew commonly describes the operation as 'turning to the right' or, 'rotating on bottom'. Almost all rigs today have a rotary table, either as primary or backup system for rotating the drillstring (see Top-drive)

Term	Application
Round trip	The complete operation of removing the drillstring from the wellbore and running it back in the hole. This operation is typically undertaken when the bit becomes dull or broken and no longer drills the rock efficiently. Once on bottom, drilling commences again. A general estimate for the round trip is one hour per 1000 feet of hole, plus an hour or two for handling collars and bits
Shaker	The primary and most important device on the rig for removing drilled solids from the mud. A wire-cloth screen vibrates while the drilling fluid flows over the top. The liquid phase of the mud and solids smaller than the wire mesh pass through the screen, whereas larger solids are retained on the screen and eventually fall off the back of the device and are collected for analysis or discarded
Side track	A secondary wellbore drilled away from the original hole. It is possible to have multiple side tracks, each of which might be drilled for a different reason
Top-drive	Top-drive technology allows continuous rotation of the drillstring and has replaced the rotary table in certain operations. The top-drive is suspended from the hook, so the rotary mechanism is free to travel up and down the derrick. A few rigs are being built today with top-drive systems only and lack the traditional kelly system
Travelling block	The set of sheaves that move up and down in the derrick. The wire rope threaded through them is threaded back to the stationary crown blocks located on the top of the derrick. This pulley system gives great mechanical advantage, enabling heavy loads (drillstring, casing and liners) to be lifted out of or lowered into the wellbore
Underbalance	The amount of pressure exerted on a formation exposed in a wellbore below the internal fluid pressure of that formation. If sufficient porosity and permeability exist, formation fluids enter the wellbore. The drilling rate typically increases as an underbalanced condition is approached
Washout	A washout in an open-hole section is larger than the original hole size or size of the drill bit. Generally, washouts become more severe with time. Appropriate mud types, mud additives and increased mud density can minimize washouts
Wiper trip	A trip made to clean the open-hole, often made before logging or between logging runs if the hole becomes unstable or "sticky"

References

Amaefule, J.O., Altunbay, M., Tiab, D., et al. (1993) Enhanced reservoir description: using core and log data to identify hydraulic (flow) units and predict permeability in uncored intervals/wells. Presented at the 68th SPE Annual Technical Conference and Exhibition, Houston, TX, paper SPE 26436.

Archie, G.E. (1942) The electrical resistivity log as an aid in determining some reservoir characteristics. *Transactions of the American Institute of Mining and Metallurgical Engineers*, **146**: 54–62.

Asquith, G.B. (1985) *Handbook of Log Evaluation Techniques for Carbonate Reservoirs*. Methods in Exploration No. 5. Tulsa, OK: American Association of Petroleum Geologists.

Asquith, G.B. and Krygowski, D. (2004) Basic Well Log Analysis. Methods in Exploration No.16 Tulsa, O.K.: American Association of Petroleum Geologists.

Bacon, M., Simm, R. and Redshaw, T. (2003) *3-D Seismic Interpretation*. Cambridge: Cambridge University Press.

Bastin, J.C., Boycott-Brown, T., Sims, A. and Woodhouse, R. (2003) The South Morecambe Gas Field, Blocks 110/2a, 110/3a, 110/7a and 110/8a, East Irish Sea. In: *United Kingdom Oil and Gas Fields, Commemorative Millennium Volume* (ed. Gluyas, J.G. and Hitchens, H.M.). Geological Society Memoir No. 20. London: Geological Society, pp. 107–18.

Bear, J. (1972) *Dynamics of Fluids in Porous Media*. New York: American Elsevier.

Cannon, S.J.C. (1994) Integrated facies description. *DiaLog*, **2**(3):4–5 (reprinted in *Advances in Petrophysics – 5 Years of DiaLog 1993–1997*. London: Petrophysical Society, 1999, pp. 7–9).

Clavier, C., Coates, G. and Dumanoir, J. (1977) The theoretical and experimental basis for the 'dual water' model for the interpretation of shaly sands. 52nd Annual Conference, SPE of AIME. *Society of Petroleum Engineers Paper SPE 6859*.

Coates, G. and Dumanoir, J.L. (1973) A new approach to improve log-derived permeability. In *Transactions of the Society of Professional Well Log Analysts 14th Annual Logging Symposium*, Paper R.

Corbett, P.W.M. and Potter, D.K. (2004) Petrotyping: a basemap and atlas for navigating through permeability and porosity data for reservoir comparison and permeability prediction. Presented at the International Symposium of the Society of Core Analysts, Abu Dhabi, SCA Papers 2004-30.

Cuddy, S.G., Allinson, G. and Steele, R. (1993) A simple convincing model for calculating water saturations in southern North Sea gas fields. In: *Transactions of the Society of Professional Well Log Analysts 34th Annual Logging Symposium, Calgary, Alberta, Canada*, pp. H1–H17.

Doll, H.G. (1948) The SP log: theoretical analysis and principles of interpretation. *Transactions of the AIME*, **179**: 146–85.

Doveton, J.H. (1994) *Geologic Log Analysis Using Computer Methods*. AAPG Computer Applications in Geology No. 2. Tulsa OK: American Association of Petroleum Geologists.

Doveton, J.H. and Cable, H.W. (1979) Fast matrix methods for the lithological interpretation of geophysical logs. In: *Petrophysical and Mathematical Studies of Sediments* (ed. Merriman, D.F.). Oxford: Pergamon Press, pp. 79–91.

Fertl, W.H. and Vercellino, W.C. (1978) Predict water cut from well logs. In *Practical Log Analysis, Part 4. Oil and Gas Journal*, 15 May 1978–19 September 1979.

Folk, R.L. (1980) *Petrology of Sedimentary Rocks*. Cedar Hill, TX: Hemphill Publishing.

Gassmann, F., (1951) Elastic waves through a packing of spheres. *Geophysics*, **16**:673–85.

Guillotte, J.G., Schrank, J. and Hunt, E. (1979) Smackover reservoir: interpretation case study of water saturation versus production. *Transactions: Gulf Coast Association of Geological Societies*, **29**:121–6.

Hilchie, D.W. (1978) *Applied Openhole Log Interpretation*. Golden, CO: D.W. Hilchie Inc.

Keith, B.D. and Pittman, E.D. (1983) Bimodal porosity in oolitic reservoirs – effect on productivity and log response, Rodessa Limestone (Lower Cretaceous), East Texas basin. *AAPG Bulletin*, **67**:1391–9.

Kenyon, W.E., Day, P.I., Straley, C., et al. (1988) A three-part study of NMR longitudinal relaxation properties of water-saturated sandstones. *Society of Petroleum Engineers Formation Evaluation*, **3**: 622–36.

Leverett, M.C. (1941) Capillary behaviour in porous solids. *Transactions of the AIME*, **142**:159–72.

Lucia, F.J. (1983) Petrophysical parameters estimated from visual description of carbonate rocks: a field classification of carbonate pore space. *Journal of Petroleum Technology*, **35**:626–37.

Lucia, F.J. (1999) *Carbonate Reservoir Characterization*. Berlin: Springer.

Lucia, F.J. and Conti, R.D. (1987) *Rock Fabric, Permeability and Log Relationships in an Upward-Shoaling, Vuggy Carbonate Sequence*. Geological Circular No. 87–5. Austin, TX: Bureau of Economic Geology, University of Texas at Austin.

Mayer-Gürr, A. (1976) *Petroleum Engineering. Geological Prospecting of Petroleum (Geology of Petroleum)* (ed. Beckmann, H.), vol. **3**. Stuttgart: Ferdinand Enke Verlag.

Nugent, W.H., Coates, G.R. and Peebler, R.P. (1978) A new approach to carbonate analysis. In *Transactions of the Society of Professional Well Log Analysts 19th Annual Logging Symposium*, Paper O.

Nurmi, R.D. (1984) Carbonate pore systems: porosity/permeability relationships and geological analysis (abstract). Presented at the AAPG Annual Meeting, San Antonio, TX, 20–23 May.

Passey, Q.R., Creaney, S., Kulla, J.B., et al. (1990) A practical model for organic richness from porosity and resistivity logs. *AAPG Bulletin*, **74**:1777–94.

Quirein, J.A., Garden, J.S. and Watson, J.T. (1982) Combined natural gamma ray spectral/litho-density measurements applied to complex lithologies. *SPE Paper 11143*, I–14.

Rider, M.H. and Kennedy, M.C. (2011) *The Geological Interpretation of Well Logs*, 3rd edn. Rogart, Sutherland: Rider-French Consulting.

Ringrose, P.S. (2008) Total-property modelling: dispelling the net-to-gross myth. *SPE Reservoir Evaluation and Engineering*, **11**:866–73.

Ringrose, P.S. and Bentley, M.R. (2014) *Reservoir Model Design. A Practitioner's Guide*. Dordrecht: Springer Science+Business Media.

Schmoker, J.W. (1981) Organic-matter content of Appalachian Devonian shales determined by use of wire-line logs. *AAPG Bulletin*, **63**:1504–37.

Schroeder, F.W. (2006) *Well-Seismic Ties*. AAPG Visiting Geoscientist, Lecture 7, Lecture Slides 1–19. Tulsa, OK: American Association of Petroleum Geologists.

Shaker, S. (2007) Calibration of geopressure predictions using normal compaction trend: perception and pitfall. *CSEG Recorder*, January:29–30.

Simandoux, P. (1963) Dielectric measurements on porous media application to the measurement of water saturations: study of the behaviour of argillaceous formations. *Revue de l'Institut Français du Petrole*, **18**(Suppl.): 193–215.

Skelt, C. and Harrison, B. (1995) An integrated approach to saturation height analysis. In: *Transactions of the SPWLA 36th Well Annual Logging Symposium, Paris, France*, paper NNN.

Teeuw, D. (1971) Prediction of formation compaction from laboratory compressibility data. *Society of Petroleum Engineers Journal*, **11**: 263–71.

Tiab, D. and Donaldson, E.C. (1996) *Petrophysics; Theory and Practice of Measuring Reservoir Rock and Fluid Transport Properties*. Houston, TX: Gulf Publishing.

Timur, A. (1968) An investigation of permeability, porosity, and residual water saturation relationships for sandstone reservoirs. *The Log Analyst*, **9**(4): 8–17.

Tittman, J. and Wahl, J.S. (1965) The physical foundations of formation density logging (gamma-gamma), *Geophysics*, **30**: 284–94.

Waxman, M.H. and Smits, L.J.M. (1968) Electrical conductivities in oil-bearing shaly sands. *SPE Journal*, **8**: 107–22.

Winsauer, W.O., Shearing, H.M., Jr, Masson, P.H. and Williams, M. (1952) Resistivity of brine saturated sands in relation to pore geometry. *AAPG Bulletin*, **36**: 253–77.

Worthington, P.F. (1985) The evolution of shaly-sand concepts in reservoir evaluation. *The Log Analyst*, **26**: 23–40.

Worthington, P.F. (2002) Application of saturation-height functions in integrated reservoir description. In: Geological Application of Well Logs (ed. Lovell, M. and Parkinson, N.). AAPG Methods in Exploration No. 13. Tulsa, OK: American Association of Petroleum Geologists, pp. 75–89.

Index

Petrophysics: A Practical Guide, First Edition. Steve Cannon.
© 2016 John Wiley & Sons, Ltd. Published 2016 by John Wiley & Sons, Ltd.